# AQA
## A-level

# Biology

(1)

Pauline Lowrie, Mark Smith
Mike Bailey, Bill Indge
Martin Rowland

**Approval message from AQA**

This textbook has been approved by AQA for use with our qualification. This means that we have checked that it broadly covers the specification and we are satisfied with the overall quality. Full details of our approval process can be found on our website.

We approve textbooks because we know how important it is for teachers and students to have the right resources to support their teaching and learning. Please note, however, that the publisher is ultimately responsible for the editorial control and quality of this book.

Please note that when teaching the *AQA A-level Biology* course, you must refer to AQA's specification as your definitive source of information. While this book has been written to match the specification, it cannot provide complete coverage of every aspect of the course.

A wide range of other useful resources can be found on the relevant subject pages of our website: www.aqa.org.uk.

DYNAMIC LEARNING

HODDER
EDUCATION
AN HACHETTE UK COMPANY

Although every effort has been made to ensure that website addresses are correct at time of going to press, Hodder Education cannot be held responsible for the content of any website mentioned in this book. It is sometimes possible to find a relocated web page by typing in the address of the home page for a website in the URL window of your browser.

Hachette UK's policy is to use papers that are natural, renewable and recyclable products and made from wood grown in sustainable forests. The logging and manufacturing processes are expected to conform to the environmental regulations of the country of origin.

Orders: please contact Bookpoint Ltd, 130 Milton Park, Abingdon, Oxon OX14 4SB. Telephone: +44 (0)1235 827720. Fax: +44 (0)1235 400454. Lines are open 9.00a.m.–5.00p.m., Monday to Saturday, with a 24-hour message answering service. Visit our website at www.hoddereducation.co.uk

© Pauline Lowrie, Mark Smith and Mike Bailey, Bill Indge, Martin Rowland 2015

First published in 2015 by
Hodder Education,
An Hachette UK Company
338 Euston Road
London NW1 3BH

Impression number    10 9 8 7 6 5 4 3 2 1

Year                2019  2018  2017  2016  2015

Cover photo © Sebastian Duda – Fotolia
Typeset in 11/13 pt ITC Berkeley Oldstyle Std by Aptara, Inc.
Printed in Italy

A catalogue record for this title is available from the British Library

**ISBN 9781471807619**

# Contents

# Get the most from this book

Welcome to the **AQA A-level Biology Year 1 Student's Book**. This book covers Year 1 of the AQA A-level Biology specification and all content for the AQA AS Biology specification.

The following features have been included to help you get the most from this book.

## Prior knowledge

This is a short list of topics that you should be familiar with before starting a chapter. The questions will help to test your understanding.

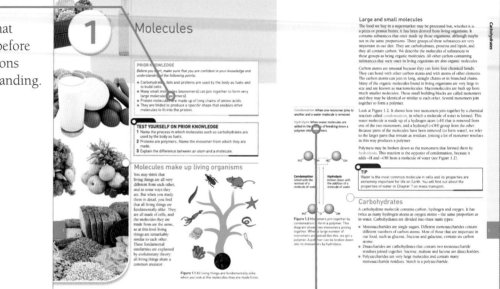

## Key terms and formulae

These are highlighted in the text and definitions are given in the margin to help you pick out and learn these important concepts.

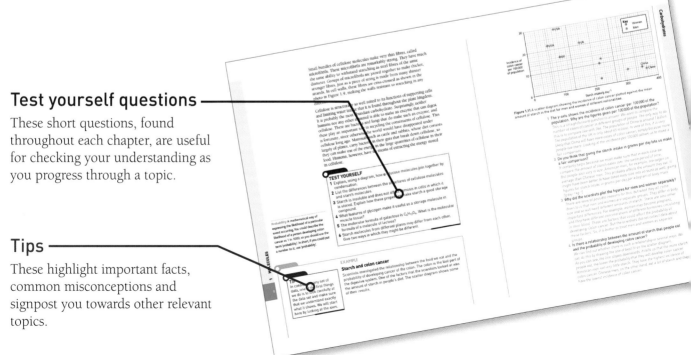

## Test yourself questions

These short questions, found throughout each chapter, are useful for checking your understanding as you progress through a topic.

## Tips

These highlight important facts, common misconceptions and signpost you towards other relevant topics.

## Activities and Required practicals

These practical-based activities will help consolidate your learning and test your practical skills. AQA's required practicals are clearly highlighted.

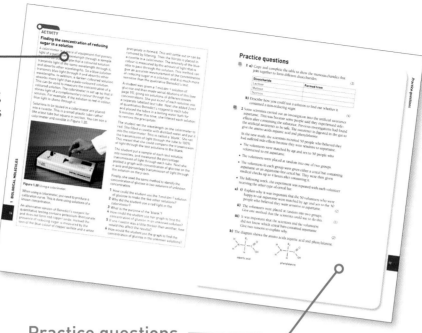

## Practice questions

You will find Practice questions at the end of every chapter. These follow the style of the different types of questions you might see in your examination, including multiple-choice questions, and are colour coded to highlight the level of difficulty. Test your understanding even further, with Maths questions and Stretch and challenge questions.

- Green – Basic questions that everyone should be able to answer without difficulty.
- Orange – Questions that are a regular feature of exams and that all competent candidates should be able to handle.
- Purple – More demanding questions which the best candidates should be able to do.
- Stretch and challenge – Questions for the most able candidates to test their full understanding and sometimes their ability to use ideas in a novel situation.

## Extension

Throughout the book you will also find Extension boxes, which contain extra material to deepen your understanding of a topic.

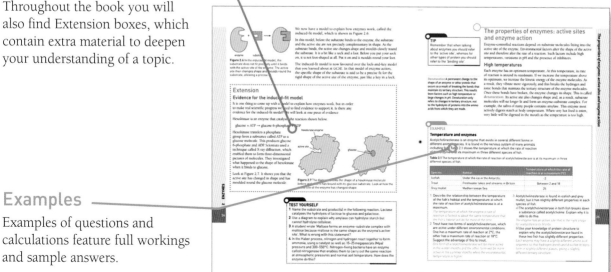

## Examples

Examples of questions and calculations feature full workings and sample answers.

Dedicated chapters for developing your **Maths** and **Practical skills** and **Preparing for your exam** can be found at the back of this book.

# 1 Biological molecules

## PRIOR KNOWLEDGE

*Before you start, make sure that you are confident in your knowledge and understanding of the following points.*

- Carbohydrates, fats and proteins are used by the body as fuels and to build cells.
- Many small molecules (monomers) can join together to form very large molecules (polymers).
- Protein molecules are made up of long chains of amino acids.
- They are folded to produce a specific shape that enables other molecules to fit into the protein.

## TEST YOURSELF ON PRIOR KNOWLEDGE

1 Name the process in which molecules such as carbohydrates are used by the body as fuels.
2 Proteins are polymers. Name the monomer from which they are made.
3 Explain the difference between an atom and a molecule.

## Molecules make up living organisms

You may think that living things -are all very different from each other, and in some ways they are. But when you study them in detail, you find that all living things are fundamentally alike. They are all made of cells, and the molecules they are made from are the same, so at this level living things are remarkably similar to each other. These fundamental similarities are explained by evolutionary theory: it is thought that all living things share a common ancestor.

**Figure 1.1** All living things are fundamentally alike when you look at the molecules they are made from.

**Condensation** When one monomer joins to another and a water molecule is removed.

**Hydrolysis** When water molecules are added in the process of breaking bonds between molecules, for example when breaking a polymer into monomers.

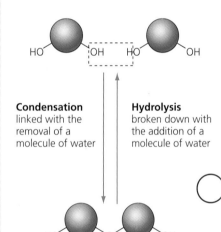

**Condensation** linked with the removal of a molecule of water

**Hydrolysis** broken down with the addition of a molecule of water

**Figure 1.2** Monomers join together by condensation to form a polymer. This diagram shows two monomers joining together. When a large number of monomers are joined like this, we get a polymer. A polymer can be broken down into its monomers by hydrolysis.

## Large and small molecules

The food we buy in a supermarket may be processed but, whether it is a pizza or peanut butter, it has been derived from living organisms. It contains substances that once made up those organisms, although maybe not in the same proportions. Three groups of these substances are very important in our diet. They are carbohydrates, proteins and lipids, and they all contain carbon. We describe the molecules of substances in these groups as being organic molecules. All other carbon-containing substances that were once in living organisms are also organic molecules.

Carbon atoms are unusual because they can form four chemical bonds. They can bond with other carbon atoms and with atoms of other elements. The carbon atoms can join in long, straight chains or in branched chains. Many of the organic molecules found in living organisms are very large in size and are known as macromolecules. Macromolecules are built up from much smaller molecules. In many macromolecules these small building blocks are called monomers and they may be identical or similar to each other. Several monomers join together to form a polymer.

Look at Figure 1.2. It shows how two monomers join together by a chemical reaction called condensation, in which a molecule of water is formed. This water molecule is made up of a hydrogen atom (–H) that is removed from one of the two monomers, and a hydroxyl (–OH) group from the other. Because parts of the molecules have been removed (to form water), we refer to the larger parts that remain as residues. Joining a lot of monomer residues in this way produces a polymer.

Polymers may be broken down to the monomers that formed them by hydrolysis. This reaction is the opposite of condensation, because it adds –H and –OH from a molecule of water (see Figure 1.2).

# Carbohydrates

A carbohydrate molecule contains carbon, hydrogen and oxygen. It has twice as many hydrogen atoms as oxygen atoms – the same proportion as in water. Carbohydrates are divided into three main types:

● Monosaccharides are single sugars. Different monosaccharides contain different numbers of carbon atoms. Most of those that are important in our food, such as glucose, fructose and galactose, contain six carbon atoms.
● Disaccharides are carbohydrates that contain two monosaccharides joined together. Sucrose, maltose and lactose are disaccharides.
● Polysaccharides are very large molecules and contain many monosaccharides. Starch is a polysaccharide.

# Glucose and other sugars

Glucose is a monosaccharide, so it is a single sugar. Its molecular formula is $C_6H_{12}O_6$. This formula simply tells us how many atoms of each element there are in each glucose molecule.

Now look at the structural formulae shown in Figure 1.4. They show a molecule of α-glucose and a molecule of β-glucose. Count each type of atom in diagram (a). There are 6 carbon atoms, 12 hydrogen atoms and 6 oxygen atoms, equal to the numbers of different atoms shown by the molecular formula, $C_6H_{12}O_6$. This diagram also shows you how the atoms are arranged.

All glucose molecules have the same formula, $C_6H_{12}O_6$. However, there are two different kinds of glucose. This is because the atoms in the glucose molecule can be arranged in different ways, called isomers. Figure 1.4 shows the arrangement of the atoms in the two different kinds of glucose.

(a)                                    (b)

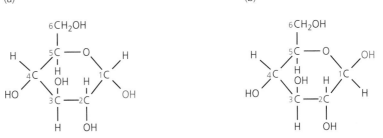

**Figure 1.4** (a) An α-glucose molecule; and (b) a β-glucose molecule.

Look at the way that the –H and –OH groups are bonded to the carbon atom on the right-hand side (1C) in β-glucose. Now look at the –H and –OH groups bonded to the carbon atom on the left-hand side (C4). Notice that they are bonded the opposite way round. Compare this with the diagram of α-glucose. Here, both –H groups are above the carbon atoms, and both –OH groups are below the carbon atoms.

Galactose and fructose are also monosaccharides and have exactly the same molecular formula as α-glucose. However, the atoms that make up these molecules are arranged in different ways. This means that, although all three substances are sugars, they have slightly different structures. This gives them slightly different properties.

Monosaccharides such as α-glucose are the monomers that join together to make many other carbohydrates. Two α-glucose molecules join by condensation to form a molecule of the disaccharide maltose. The bond forms between carbon 1 of one α-glucose molecule and carbon 4 of the other; such a bond is called a glycosidic bond (see Figure 1.5).

Other disaccharides form in a similar way. Lactose, for example, is the sugar found in milk. It is formed in a condensation reaction between a molecule of α-glucose and a molecule of another monosaccharide, galactose. Sucrose is formed from α-glucose and fructose.

When sugars such as α-glucose are boiled with Benedict's solution, an orange precipitate is formed because Cu(II) ions in the Benedict's solution are reduced to orange Cu(I) ions. This reaction occurs because of the way the chemical groups are arranged in such sugars. These sugars are therefore called reducing sugars. Fructose, maltose and galactose are also reducing sugars.

**Figure 1.3** Most of the carbohydrate that we eat comes from plants. This crop is sugar cane, a plant that stores sucrose in its stem. The carbohydrate stored by other food plants such as potatoes and cereals is starch.

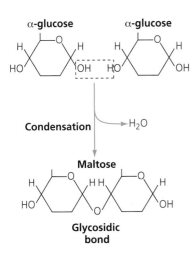

**Figure 1.5** Two α-glucose molecules join together by condensation to give a molecule of the disaccharide maltose.

**Glycosidic bond** A chemical bond formed as the result of condensation between two monosaccharides.

**Reduction** is gain of electrons or hydrogen.

# Extension

## Structure of amylose and amylopectin

Figure 1.6 shows the structure of starch. You can see that amylose is a long chain of α-glucose molecules. They are linked by 1,4-glycosidic bonds. This chain is coiled into a spiral and its coils are held in place by chemical bonds called hydrogen bonds. Amylopectin is also a polymer of α-glucose but its molecules are branched due to 1,6-glycosidic bonds.

In amylose, the α-glucose molecules are linked by 1,4-glycosidic bonds. Notice that the –CH₂OH

side-chains all stick out on the same side. This arrangement causes the chains of α-glucose molecules to coil into spirals as shown in Figure 1.6. Amylopectin molecules have branches because some of the α-glucose molecules form bonds between carbon atoms 1 and 6 instead of 1 and 4. This enables starch molecules to fold up compactly.

**Amylose**
consists of a long chain of α-glucose residues

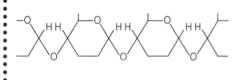

The chain is coiled into a spiral. Hydrogen bonds hold this spiral in shape.

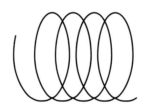

**Amylopectin**
consists of branched chains of α-glucose residues

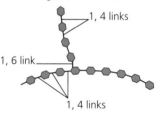

1, 4 links

1, 6 link

1, 4 links

**Figure 1.6** Starch consists of amylose and amylopectin. Starch from different plants contains different amounts of these two substances.

---

> **TIP**
> You can remember that oxidation is loss of electrons or hydrogen and reduction is gain of electrons or hydrogen by using the mnemonic OILRIG: Oxidation Is Loss, Reduction Is Gain.

Sucrose does not give an orange precipitate with Benedict's solution; it is a non-reducing sugar. However, when boiled with dilute acid, sucrose is hydrolysed to monosaccharides. The sucrose molecules are hydrolysed into α-glucose and fructose, both reducing sugars. Then it will give a positive test with Benedict's solution. (See page 14 for details of qualitative tests.)

## Starch

Starch, a substance found in plants, is one of the most important fuels in the human diet. It makes up about 30% of what we eat. Starch is a mixture of two substances, amylose and amylopectin. Both these substances are polymers made from a large number of α-glucose molecules joined together by condensation reactions. In the biochemical test for starch, you add a drop of iodine solution. A blue-black colour indicates the presence of starch.

## Storage molecules

### Starch for storage

We use the starch from plants as a fuel. For many plants, starch is a storage compound, both for short-term storage overnight when photosynthesis cannot occur, and for long-term storage, for example in seeds and in the

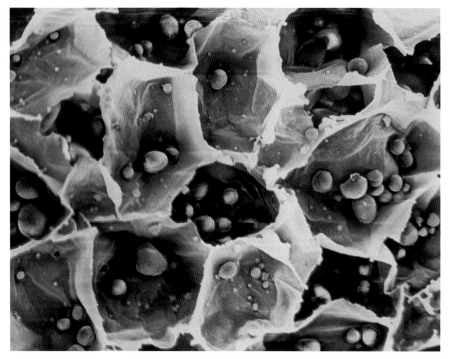

**Figure 1.7** Starch molecules can fold up compactly and can therefore fit into small storage organelles, such as the starch grains in potato tuber cells, shown here. The starch grains are shown in green.

organs such as bulbs and tubers that survive through the winter. It is particularly suited for storage because it is insoluble and so does not diffuse out of cells easily or have any effects on water potential and thus osmosis.

As a storage compound it is important that starch can be easily synthesised and broken down. Plants have enzymes that can rapidly carry out these processes.

We have a digestive enzyme called amylase that hydrolyses the starch in our diet to maltose. This can then be hydrolysed into glucose, which is needed to provide a source of fuel for respiration.

## Glycogen for storage

Animals such as humans do not rebuild excess glucose into starch for storage. Instead, we make it into a polysaccharide similar to starch called glycogen.

Like amylopectin, glycogen also consists of α-glucose chains with both 1,4- and 1,6-glycosidic bonds, but the 1,6 bonds are much more frequent, so the molecules are much more branched (see Figure 1.8, overleaf). This makes glycogen molecules even more compact than starch molecules. In humans, some glycogen is stored in the muscles as a readily accessible store of glucose close to the site where the rate of respiration is regularly raised very rapidly. The liver stores larger reserves of glycogen and continually breaks it down to maintain a stable blood glucose concentration.

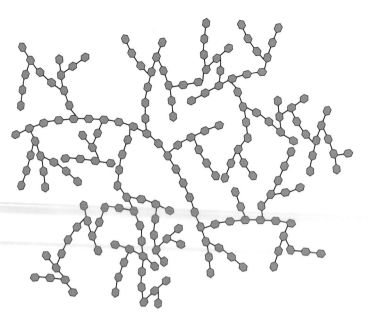

**Figure 1.8** A glycogen molecule.

## Cellulose for strength

The main substance in a plant cell wall is the carbohydrate cellulose. Like starch, cellulose is a polysaccharide and is a polymer of glucose. The monomer in cellulose is β-glucose.

**Hydrogen bond** A chemical bond important in the three-dimensional structure of biological molecules. Hydrogen bonds require relatively little energy to break.

In cellulose, the β-glucose molecules join together in chains by condensation. As when starch chains are made from α-glucose molecules, glycosidic bonds are formed. But in the cellulose chains, every other β-glucose is 'upside-down', so the –CH₂OH side-chains stick out alternately on opposite sides, as you can see in Figure 1.9. This 'alternate' bonding makes the cellulose molecules very straight. They are also very long. They line up parallel with each other and become linked together by many hydrogen bonds. Although each hydrogen bond is weak, many together lead to strong binding between the molecules.

**Figure 1.9** Cellulose is a polymer of β-glucose molecules joined by glycosidic bonds. Its molecules are long and straight and form fibres that are very strong. Cellulose gives cell walls their strength and resistance to being stretched.

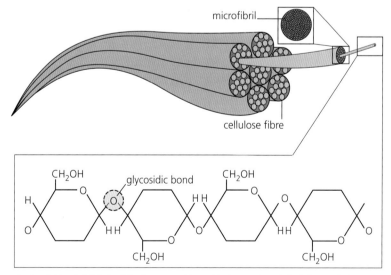

Long chain of 1,4 linked β-glucose residues. Hydrogen bonds link these chains together to form microfibrils.

Small bundles of cellulose molecules make very thin fibres, called microfibrils. These microfibrils are remarkably strong. They have the ability to withstand stretching as steel fibres of the same diameter. Groups of microfibrils are joined together to make thicker, stronger fibres, just as a piece of string is made from many thinner strands. In cell walls, these fibres are criss-crossed, making the walls resistant to stretching in any direction.

Cellulose is structurally so well suited to its functions of supporting cells and limiting water intake that it is found throughout the plant kingdom. It is probably the most abundant carbohydrate. Surprisingly, neither humans nor any other mammal are able to make an enzyme that can digest cellulose. There are bacteria and fungi that do make such an enzyme, and these play an important role in recycling the constituents of cellulose. This is fortunate, since otherwise the world would have disappeared under cellulose long ago. Mammals such as cattle and rabbits, whose diet consists largely of plants, carry bacteria in their guts that hydrolyse cellulose, so they can make use of the energy in the large quantities of cellulose in their food. Humans, however, have no means of extracting the energy stored in cellulose.

**TEST YOURSELF**

1 Explain, using a diagram, how α-glucose molecules join together by condensation.
2 List the differences between the structures of cellulose molecules and starch molecules.
3 Starch is insoluble and does not affect osmosis in cells in which it is stored. Explain how these properties make starch a good storage compound.
4 What features of glycogen make it useful as a storage molecule in muscle tissue?
5 The molecular formula of galactose is $C_6H_{12}O_6$. What is the molecular formula of a molecule of lactose?
6 Starch molecules from different plants may differ from each other. Give two ways in which they might be different.

# Lipids

The term 'lipids' covers a group of substances that includes fats and oils (triglycerides), steroids and sterols, and waxes. Two groups of lipids are especially significant. These are triglycerides and phospholipids. You will learn in Chapter 3 that the cell-surface membrane is made up of lipids and proteins.

## Triglycerides

The commonest lipids found in living organisms are triglycerides. Most of the triglycerides found in animals are known as fats. They are solid at a temperature of about 20 °C. A triglyceride is made up of a molecule of glycerol and three fatty acid molecules. The basic structures of these molecules are shown in Figure 1.10, overleaf.

As Figure 1.10 shows, there are two kinds of fatty acids. Saturated fatty acids have only single bonds between the carbon atoms. Unsaturated fatty acids have at least one double bond between carbon atoms. In general,

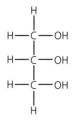

R.COOH

**(a)** Glycerol is a type of alcohol. It has three –OH groups, each of which can undergo a condensation reaction with a fatty acid.

**(b)** This is the simplest formula for a fatty acid molecule. The letter R represents a hydrocarbon chain consisting of carbon and hydrogen atoms.

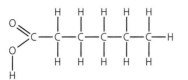

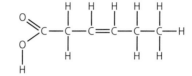

**(c)** In saturated fatty acids, each of the carbon atoms in this chain, with the exception of the last, has two hydrogen atoms joined to it. The bonds between the carbon atoms are single bonds.

**(d)** In unsaturated fatty acids, there are one or more double bonds between the carbon atoms in the chain. Because of this, some carbon atoms will be joined only to a single hydrogen atom.

**Figure 1.10** The basic structure of a molecule of (a) glycerol and (b) fatty acid; (c) shows the structure of a saturated fatty acid and (d) shows the structure of an unsaturated fatty acid.

saturated fatty acids have higher melting points than unsaturated fatty acids. Fats that are solid at room temperature, such as lard or butter, tend to have more saturated fatty acids in them, while oils that are liquid at room temperature, such as sunflower or olive oil, have more unsaturated fatty acids.

Glycerol is a type of alcohol. Look at Figure 1.10 (a). You will see that there are three –OH groups in glycerol. These groups allow the molecule to join with three fatty acids to produce a triglyceride. Figure 1.10 (b) is the simplest possible way of showing the structure of a fatty acid molecule. The letter R represents a chain of hydrogen and carbon atoms. In the fatty acids found in animal cells there are often 14 to 16 carbon atoms in this chain.

When a triglyceride is formed, a molecule of water is removed as each of the three fatty acids joins to the glycerol. You may remember that this type of chemical reaction is called condensation (see page 2). The formation of a triglyceride from glycerol and fatty acids is shown in Figure 1.11. The bond formed between the glycerol and the fatty acid is called an ester bond.

You can use the emulsion test to test for lipids such as triglycerides. Crush a little of the test material and mix it thoroughly with ethanol. Pour the resulting solution into water in a test tube. A white emulsion shows that a lipid is present.

- Draw a diagram to show a glycerol molecule.
- Draw three fatty acid molecules 'the wrong way round' next to it.

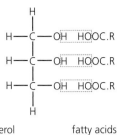

glycerol       fatty acids

**Figure 1.11** This diagram is a simple way of showing how a molecule of glycerol joins with three fatty acid molecules to form a triglyceride.

- Remove three molecules of water, taking the H from the glycerol and the –OH from the fatty acids.

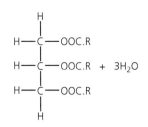

- Close everything up to show the completed triglyceride.

## Phospholipids

A phospholipid has a very similar structure to a triglyceride, but as you can see from Figure 1.12, it contains a phosphate group instead of one of the fatty acids. It is quite a good idea to think of a phospholipid as having a 'head' consisting of glycerol and phosphate and a 'tail' containing the long chains of hydrogen and carbon atoms in the two fatty acids. The presence of the phosphate group means that the 'head' is attracted to water. It is therefore described as being **hydrophilic** or 'water loving'. The hydrocarbon tails do not mix with water, so this end of the molecule is described as **hydrophobic** or 'water hating'.

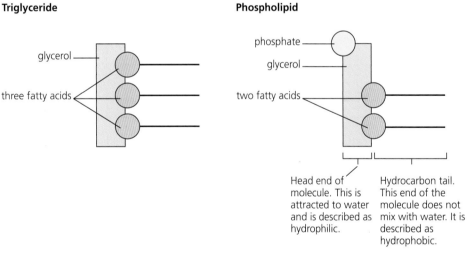

**Figure 1.12** A phospholipid has a structure very similar to a triglyceride, but it contains a phosphate group instead of one of the fatty acids.

When phospholipids are mixed with water, they arrange themselves in a double layer with their hydrophobic tails pointing inwards and their hydrophilic heads pointing outwards. This double layer is called a phospholipid bilayer and forms the basis of membranes in and around cells.

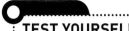

**7** Triglycerides are not polymers. Explain why.

**8** Carbohydrates and triglycerides are both made of carbon, hydrogen and oxygen atoms. Explain how the proportions of these atoms are different in carbohydrates and triglycerides.

**9** How is a triglyceride different from a phospholipid?

## ACTIVITY

### Fatty acids in milk

Milk contains triglycerides. Scientists investigated whether the fatty acids in human breast milk depend on the food that the mother eats. The scientists collected samples of breast milk from two groups of women. The women in one group were vegans and only ate food obtained from plants. Those in the other group, the control group, ate food obtained from both animals and plants. Table 1.1 shows the concentrations of different fatty acids in the milk samples.

**Table 1.1** The concentrations of different fatty acids in vegan and control group milk samples.

| Fatty acid | Number of double bonds in hydrocarbon chain | Number of carbon atoms in hydrocarbon chain | Concentration of fatty acid in breast milk sample/mg g$^{-1}$ | |
|---|---|---|---|---|
| | | | Vegan group | Control group |
| Lauric | 0 | 12 | 39 | 33 |
| Myristic | 0 | 14 | 68 | 80 |
| Palmitic | 0 | 16 | 166 | 276 |
| Stearic | 0 | 18 | 52 | 108 |
| Palmitoleic | 1 | 16 | 12 | 36 |
| Oleic | 1 | 18 | 313 | 353 |
| Linoleic | 2 | 18 | 317 | 69 |
| Linolenic | 3 | 18 | 15 | 8 |

**1** The first four fatty acids in the table are saturated fatty acids. Explain why they are described as saturated.

**2** Construct a table to show all of the following:
- the total concentration of saturated fatty acids in milk from the vegan group
- the total concentration of unsaturated fatty acids in milk from the vegan group
- the total concentration of saturated fatty acids in milk from the control group
- the total concentration of unsaturated fatty acids in milk from the control group.

**3** Use an example from the table to explain what is meant by a polyunsaturated fatty acid.

**4** Describe the difference between the total concentration of polyunsaturated fatty acids in milk produced by the vegan group and by the control group. Suggest an explanation for this difference.

## Proteins

Earlier in this chapter, we saw that starch is a polymer made up of a single type of monomer, α-glucose. Whether these α-glucose monomers are joined to form straight chains or branched chains, they still form starch. Different types of starch are very similar.

Proteins are different. The basic building blocks of proteins are amino acids. There are 20 different amino acids found in almost all living organisms, which is indirect evidence for evolution. These amino acids can be joined in a range of different orders. In any living organism, there are a huge number of different proteins and they have many different functions.

If we take a single tissue, such as blood, we can get some idea of just how varied and important are the roles of proteins. Human blood is red because it contains haemoglobin. This is an iron-containing protein that plays an extremely important part in transporting oxygen from the lungs to respiring cells. When you cut yourself, blood soon clots. This is because another protein, fibrin, forms a mesh of threads over the surface of the wound, trapping red blood cells and forming a scab. Blood also contains enzymes, which are proteins. The antibodies produced by white blood cells are also proteins, and are important in protecting the body against disease.

The biuret reaction enables us to test for a protein. Sodium hydroxide solution is added to a test sample, and then a few drops of dilute copper sulfate solution. If there is a protein present, the solution will turn mauve.

## Amino acids: the building blocks of proteins

All 20 amino acids have the same general structure. Look at Figure 1.13. Notice that there is a central carbon atom called the α-carbon and that it is attached to four groups of atoms. There is an amine group (–$NH_2$). This is the group that gives the molecule its name. Then we have a carboxyl group (the acid; –COOH) and a hydrogen atom (–H). These three features are exactly the same in all 20 amino acids. The fourth group, called the R-group, differs from one amino acid to another. As well as showing the general structure of an amino acid, Figure 1.13 also shows the structures of three particular amino acids found in proteins. In each of these three amino acids (and in the other 17), it is only the R-group that is different.

**Peptide bond** A chemical bond formed between two amino acids as a result of condensation.

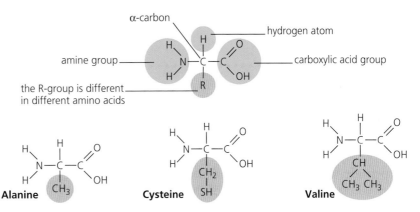

**Figure 1.13** The structure of amino acids.

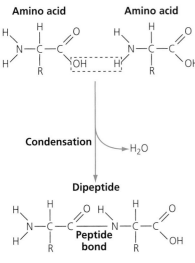

**Figure 1.14** Joining amino acids.

Amino acids join together by condensation reactions. Look at Figure 1.14. You can see that a hydrogen atom is removed from the amino group of one amino acid. This combines with an –OH group removed from the carboxylic acid of the other amino acid, forming a molecule of water. The bond formed between the two amino acid residues is called a peptide bond. Joining two amino acids together produces a dipeptide. When many amino acids are joined in this way, they form an unbranched chain called a polypeptide. Polypeptides can be broken down again by hydrolysis into the amino acids from which they are made.

## Polypeptides and proteins

**Primary structure** The sequence of amino acids in a polypeptide.

A protein consists of one or more polypeptide chains folded into a complex three-dimensional shape. Different proteins have different shapes, determined by the order in which the amino acids are arranged in the polypeptide chains. The sequence of amino acids in a polypeptide chain is the primary structure of a protein (see Figure 1.15).

**Figure 1.15** This diagram shows the primary structure of an enzyme called ribonuclease. The names of the amino acids that make up this protein have been abbreviated. Ribonuclease has 124 amino acids. Some proteins, such as antibodies, are much larger and contain many more amino acids.

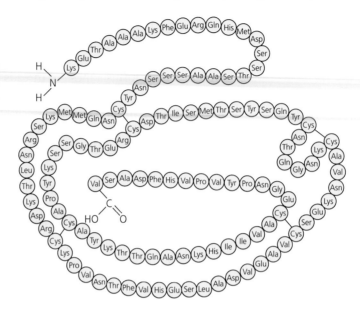

We shall see in Chapter 4 that genes carry the genetic information that enables cells to make polypeptides and ensures that the sequence of amino acids is the same in all molecules of a particular polypeptide. Changing a single one of these amino acids may be enough to cause a change in the shape of the protein and prevent it from carrying out its normal function.

**Secondary structure** The shape the polypeptide chain folds into, such as an alpha helix or a beta pleated sheet.

Parts of a polypeptide chain are twisted and folded. This is the secondary structure of a protein.

# Extension

## Secondary structure: α-helix and β-pleated sheet

Sometimes the chain, or part of it, coils to produce a spiral or α-helix. Other parts of the polypeptide may form a β-pleated sheet; this occurs where two or more parts of the chain run parallel to each other and are linked to each other by hydrogen bonds.

The sequence of amino acids in the polypeptide decides whether an α-helix or a β-pleated sheet is formed. Some sequences are more likely to form an α-helix, while others form a β-pleated sheet, as in Figure 1.16.

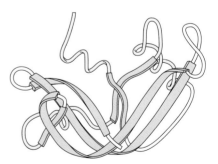

**Figure 1.16** Here is another diagram of a ribonuclease molecule, this time showing its secondary structure. The three spiral yellow parts of the polypeptide chain are where it is coiled into an α-helix. The flat blue sections show where the chain is folded to form a β-pleated sheet.

**Tertiary structure** Gives a protein the characteristic complex, three-dimensional shape that is closely related to its function.

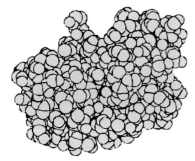

**Figure 1.17** The model in this diagram shows the tertiary structure of a ribonuclease molecule. (The shapes represent atoms.) This is the way the whole polypeptide is folded.

**TIP**
Enzymes are proteins. You will learn more about them in Chapter 2.

**(a)**

**(b)**

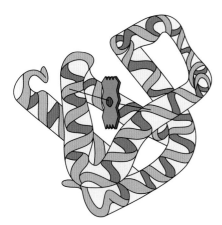

**Figure 1.18** (a) Fibrous and (b) globular proteins.

The twisted and folded chain may fold up further to give the whole polypeptide molecule a globular shape. The complex folding of the whole molecule is the tertiary structure of the protein (Figure 1.17).

As with the secondary structure, the tertiary structure is also determined by the sequence of amino acids in the polypeptide chain. All molecules of a particular protein have the same sequence of amino acids, so under the same conditions they will all fold in the same way to produce molecules with the same tertiary structure.

Tertiary structure is extremely important and is very closely related to the function of the protein. Different types of bond form between different amino acids within the protein and the types of bond help to maintain the shape of the protein. These bonds include the following:

● Hydrogen bonds (see page 6), which form between the R-groups of a variety of amino acids. These bonds are not strong. They are easily broken, but there are many of them.
● Ionic bonds, which form between an amino acid with a positive charge and an amino acid with a negative charge, if they are close enough to each other. These are not strong bonds and are easily broken.
● Disulfide bridges, which form between amino acids that contain sulfur in their R-groups. These are quite strong covalent bonds, less easily broken than hydrogen bonds or ionic bonds.

There are two categories of proteins, differing in their tertiary structure. Fibrous proteins are typically long and thin, and they are insoluble. They often have structural functions, such as keratin in hair or collagen that makes up a lot of connective tissue in our bodies. Globular proteins are more spherical in shape. They are soluble and have biochemical functions, such as enzymes or myoglobin, a pigment that stores oxygen in muscle tissue.

Some proteins have more than one polypeptide chain. We describe a protein that is made up from two or more polypeptide chains as having a quaternary structure. The polypeptide chains are held together by the same sorts of chemical bond that maintain the tertiary structure. The ribonuclease molecule shown in Figure 1.17 does not have a quaternary structure because it consists of only one polypeptide chain. The red pigment in our blood, haemoglobin, is a protein that does have a quaternary structure. A molecule of human haemoglobin has four polypeptide chains.

**TEST YOURSELF**
**10** Polypeptides can be made up from 20 different amino acids. A tripeptide is a polypeptide consisting of three amino acids. How many different tripeptides is it possible to make?
**11** Give one way in which the formation of a peptide bond is similar to the formation of a glycosidic bond.
**12** Egg white contains a protein. Which one (or more) of the following occurs when egg white is heated in a water bath containing water at 100 °C?
  **A** Glycosidic bonds are broken.
  **B** The protein is killed by the heat.
  **C** The bonds holding the tertiary structure are broken.
  **D** The protein is hydrolysed.

# Qualitative tests for substances in food

**TIP**

Qualitative tests detect the presence of a substance but do not show exactly how much is present.

There are some tests that can be carried out to find out which substances are present in samples of food and other substances. These tests are summarised in Table 1.2.

**Table 1.2** Tests for food substances.

| Substance | Test | Brief details of test | Positive result |
|---|---|---|---|
| Protein | Biuret test | • Add sodium hydroxide to the test sample. <br>• Add a few drops of dilute copper sulfate solution. | Solution turns mauve |
| Carbohydrates Reducing sugars | Benedict's test | • Heat test sample with Benedict's reagent. | Orange-red precipitate is formed |
| Non-reducing sugars | | • Check that there is no reducing sugar present by heating part of the sample with Benedict's solution. <br>• Hydrolyse rest of sample by heating with dilute hydrochloric acid. <br>• Neutralise by adding sodium hydrogencarbonate. <br>• Test sample with Benedict's solution. | Orange-red precipitate is formed |
| Starch | Iodine test | • Add iodine solution. | Turns blue-black |
| Lipid | Emulsion test | • Dissolve the test sample by shaking with ethanol. <br>• Pour the resulting solution into water in a test tube. | A white emulsion is formed |

Care should be taken with the chemicals and methods described above. Consult the CLEAPSS Hazcards for each chemical, and ensure sufficient safety precautions are taken with chemicals and hot water.

**ACTIVITY**

A student was given three tubes. The table shows the contents of each tube.

| Tube | Contents of tube |
|---|---|
| A | Protein and protease (an enzyme that digests protein) that have been left together at room temperature for an hour. |
| B | Sucrose, starch, lipid |
| C | Glucose, protein |

The student carried out tests for food substances on all three tubes.
The table below shows the results she obtained.

**1** Copy the table and complete the left-hand column to indicate which tube gave which results.

| Tube | Benedict's test for reducing sugars | Benedict's test for non-reducing sugars | Iodine test for starch | Emulsion test for lipids | Biuret test for protein |
|---|---|---|---|---|---|
| | Orange-red precipitate formed | Test not carried out | Stayed yellowy-brown | Stayed clear | Mauve |
| | Stayed blue | Stayed blue | Stayed yellowy-brown | Stayed clear | Mauve |
| | Stayed blue | Orange-red precipitate formed | Blue-black colour | Milky-white emulsion formed | Stayed blue |

**2** Why did the student decide not to carry out the non-reducing sugar test on one of the tubes?

**3** Explain why two tubes gave a positive result for the biuret test.

**4** Suggest how the student could use qualitative tests to distinguish between three solutions containing different concentrations of glucose.

**1 BIOLOGICAL MOLECULES**

14

## ACTIVITY

A student decided to investigate the sensitivity of the iodine test for starch.
She was given a 1% starch solution. She used this to make serial dilutions
as shown in Figure 1.19.

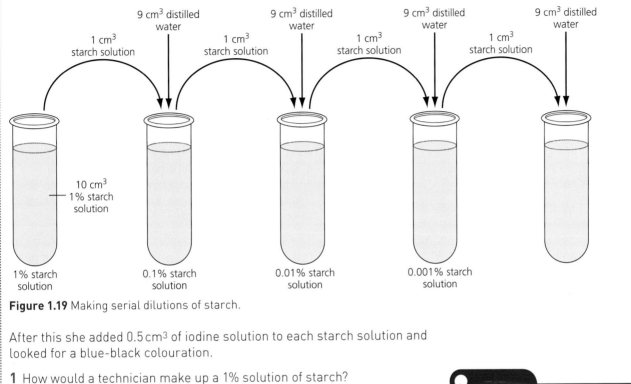

**Figure 1.19** Making serial dilutions of starch.

After this she added 0.5 cm³ of iodine solution to each starch solution and
looked for a blue-black colouration.

1 How would a technician make up a 1% solution of starch?
2 Give the concentration of the final solution in the diagram.
3 The blue-black colouration that occurs when iodine interacts with starch
  is caused by the iodine molecules becoming trapped in the amylose helix.
  Use this information to explain why the sensitivity of the iodine test can
  be different depending on the type of starch present.

**TIP**
You do not need to be able
to recall the details of this
practical activity.

## Finding the concentration of reducing sugar in a solution

A colorimeter is a piece of equipment that passes light of a particular wavelength through a sample. It works on the principle that a coloured solution transmits light of the same wavelength through it, and absorbs other wavelengths. So a blue solution transmits blue light through it and absorbs other wavelengths. In addition, a darker-coloured solution absorbs more light than a pale-coloured solution. This can be used to measure the concentration of a coloured solution. The colorimeter is set up so that it shines light of a complementary colour through the solution. For example, if the solution is red in colour, blue light is shone through it.

Solutions to be tested in a colorimeter are placed into a cuvette. This is a small plastic tube rather like a test tube but square in section. You can see a colorimeter and cuvette in Figure 1.20.

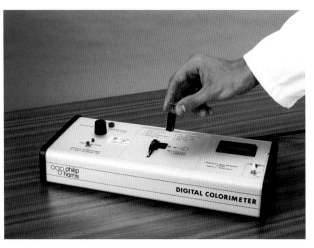

**Figure 1.20** Using a colorimeter.

When using a colorimeter, you need to produce a calibration curve. This is done using solutions of a known concentration.

An alternative version of Benedict's reagent for quantitative testing contains potassium thiocyanate and does not form red copper oxide. Instead the presence of reducing sugar is measured by the loss of the blue colour of copper sulfate and a white precipitate is formed. This will settle out or can be removed by filtering. Then the filtrate is placed in a cuvette in a colorimeter. The intensity of the blue colour is measured by the amount of light that is able to pass through the solution. This method can give an accurate measurement of the concentration of reducing sugar in a solution, and it is much more sensitive than the qualitative Benedict's test.

A student was given a $1 \, mol \, dm^{-3}$ solution of glucose and then made serial dilutions of this (see page 15), giving six solutions of different known concentrations. She put $4 \, cm^3$ of each solution into a separate labelled test tube. Next she added $2 \, cm^3$ of quantitative Benedict's reagent to each tube and placed the tubes in a boiling water bath for 5 minutes. After this time, she filtered each solution to remove the precipitate.

The student set the wavelength on the colorimeter to red. She filled a cuvette with distilled water and put it into the colorimeter. This is called a 'blank'. She set the transmission of light through the tube to 100%. This meant that she could compare the transmission of light through the test solutions to the blank.

The student put a sample of each test solution into cuvettes, and measured the percentage transmission of light through each tube. Next she plotted a graph with concentration of glucose on the $x$-axis and percentage transmission of light through the solution on the $y$-axis.

Finally, she used the same method to identify the concentration of glucose in two solutions of unknown concentration.

1 How could the student use the $1 \, mol \, dm^{-3}$ solution of glucose to make the five other solutions?
2 Why did the student use a red light in the colorimeter?
3 What is the purpose of the 'blank'?
4 How could the student use her graph to find the concentration of glucose in an unknown solution?
5 If one cuvette was a little thicker than another, how would this affect the results?
6 How would the student use the graph to find the concentration of glucose in the unknown solutions?

# Practice questions

**1 a)** Copy and complete the table to show the monosaccharides that join together to form different disaccharides. (3)

| Disaccharide | Formed from |
|---|---|
| Lactose | |
| Maltose | |
| Sucrose | |

**b)** Describe how you could test a solution to find out whether it contained a non-reducing sugar. (4)

**2** Some scientists carried out an investigation into the artificial sweetener aspartame. This was because some people said they experienced side-effects after consuming the substance. Previous investigations had found the artificial sweetener to be safe. The sweetener is digested in the gut to give the amino acids aspartic acid and phenylalanine.

In the new study, the scientists recruited 50 people who believed they had suffered side-effects because they were sensitive to aspartame.

- The volunteers were matched by age and sex to 50 people who volunteered to eat aspartame.

- The volunteers were placed at random into one of two groups.

- The volunteers in each group were given either a cereal bar containing aspartame or an aspartame-free cereal bar. They were then given medical checks up to 4 hours after consuming it.

- The following week, the experiment was repeated with each volunteer receiving the other type of cereal bar.

**a) i)** Explain why it was important that the 50 volunteers who were happy to eat aspartame were matched by age and sex to the 50 people who believed they were sensitive to aspartame. (2)

**ii)** The volunteers were placed at random into two groups. Give one method that the scientists could use to do this. (1)

**iii)** It was important that the scientists and the volunteers did not know which cereal bars contained aspartame. Give two reasons to explain why. (2)

**b)** The diagram shows the amino acids aspartic acid and phenylalanine.

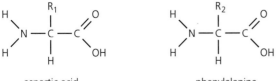

aspartic acid                     phenylalanine

Aspartame is made by joining these two amino acids together.

**i)** Draw a molecule of aspartame. (2)

**ii)** Name the reaction that occurs when these two amino acids are joined together. (1)

**3** The diagram shows two fatty acids.

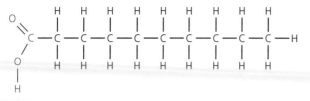

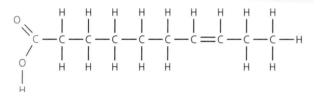

**a) i)** Which of these fatty acids is saturated? Explain your answer. (1)

**ii)** Name the reaction involved when three fatty acids combine with a glycerol molecule to form a triglyceride. (1)

**b)** Describe a test you could perform to show that a mixture contains lipids. (2)

**c)** Give one similarity and one difference between the structure of a phospholipid and the structure of a triglyceride. (2)

**4 a)** Copy and complete the table below with a tick if the statement is true and a cross if it is not true. (3)

| Statement | Proteins | Polysaccharides | Lipids |
|---|---|---|---|
| Molecule is a polymer | | | |
| Contains amino acids | | | |
| Contains nitrogen atoms | | | |

**b) i)** A protein has a tertiary structure but not a quaternary structure. How many polypeptides does it contain? (1)

**ii)** Name two kinds of bond that hold a protein in its tertiary structure. (2)

## Stretch and challenge

**5** Research the differences between D- and L-isomers of molecules. You should find that the **amino acids** in most living organisms are the **L-isomers**, and the **monosaccharides** in most living organisms are the **D-isomers**. Suggest why this is indirect evidence for evolution.

# 2 Enzymes

## Introduction

There are many chemical reactions taking place in living organisms, mainly inside the cells. The sum of all these reactions is known as **metabolism**. Some of these reactions hydrolyse larger molecules into smaller ones, while others join smaller molecules together to make bigger ones. There are enzymes in biological washing powder, which help to clean clothes at lower temperatures.

Molecules known as enzymes are required for metabolic reactions to take place. Enzymes are **biological catalysts**. They speed up the rate of chemical reactions, enabling reactions that take place in living organisms to occur fast enough for life to continue. As they are catalysts, and so remain unchanged during reactions, they can work many times.

Enzymes are made of protein and, as you will see, they have specific shapes that enable them to function. You learned in Chapter 1 (page 13) that globular proteins have a complex tertiary structure, and this is very important in explaining how enzymes work.

Enzyme A protein that speeds up the rate of a chemical reaction in a living organism. An enzyme acts as catalyst for specific chemical reactions, converting a specific set of reactants (called substrates) into specific products.

# How do enzymes work?

Have you ever found an old newspaper, one that was perhaps months or even years old? If you have, you probably noticed that it had turned a yellow-brown colour. What had happened? The paper had reacted with oxygen in the air, but it is a very, very slow reaction.

We can easily speed up this reaction. All we need to do is to touch the corner of the paper with a lighted match. The paper bursts into flame, reacting with oxygen in the air much more quickly. This reaction involves combustion. The newspaper is fuel and contains chemical potential energy. In order to release this energy in the chemical reaction with oxygen, we must supply some energy at the start. This is where the match comes in. It provides the activation energy necessary to start the reaction.

We sometimes compare what happens here with what would happen if you had a large rock at the top of a steep hill. There is a lot of potential energy in this situation but, under normal conditions, the rock will just sit there. However, give it a push and it will roll all the way down to the bottom of the hill. In other words, supply activation energy at the start, and the rock will give up a lot of its potential energy. Look at Figure 2.1. This is a graph showing, in a different way, the idea of the energy changes that take place as a chemical reaction progresses.

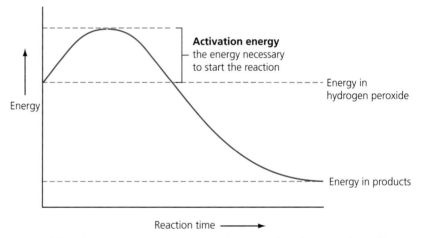

**Figure 2.1** Activation energy is necessary before a chemical reaction will take place.

Hydrogen peroxide is a substance produced by reactions in many living cells. It is harmful, so it is removed. It breaks down very slowly to give the products water and oxygen:

$$2H_2O_2 \rightarrow 2H_2O + O_2$$

The products of this reaction, water and oxygen, are not only harmless, they are extremely useful to the organism.

We can pour some hydrogen peroxide into a test tube and it will break down slowly. We can make the reaction go faster by heating the hydrogen peroxide. Clearly, inside the body we cannot use a match or a Bunsen burner to heat our cells to increase the rate at which the hydrogen peroxide they produce is broken down. This is where enzymes come in.

Cells produce an enzyme called catalase. Its **substrate**, the substance on which an enzyme acts, is hydrogen peroxide. Catalase lowers the activation energy needed to start the breakdown reaction of hydrogen peroxide. As a consequence, the hydrogen peroxide breaks down rapidly at the relatively low temperatures found inside living cells. Figure 2.2 shows this as a graph. Note that there is the same chemical potential energy at the start of the two reactions and there is the same amount in the products of both reactions. The enzyme has simply lowered the activation energy necessary to start the reaction.

**Figure 2.2** Adding an enzyme lowers the activation energy necessary to start a chemical reaction.

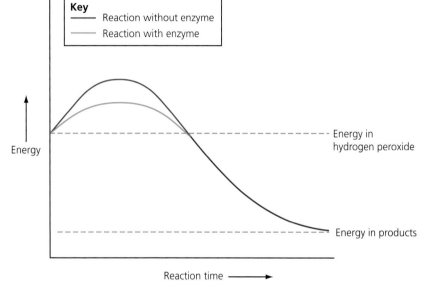

## Extension

### Enzyme shape and enzyme function: ribonuclease

A skill that a biologist must have is to be able to interpret unfamiliar data. Here we will look at some research carried out by Christian Anfinsen in the 1960s. It was very important research because it demonstrated the relationship between the structure of an enzyme and its function.

Anfinsen investigated the enzyme ribonuclease. This enzyme hydrolyses ribonucleic acid (RNA). There are different forms of ribonuclease. Look at the diagram in Figure 2.3, which shows the structure of one form of this enzyme.

The numbers on the diagram show positions along the polypeptide chain. For example, number 26 is the twenty-sixth amino acid along the chain. The large shaded dots represent the amino acid cysteine. This amino acid contains sulfur. Chemical bonds called disulfide bonds, or disulfide bridges (see Chapter 1, page 13), form between sulfur-containing

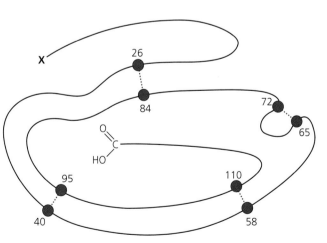

**Figure 2.3** A molecule of one form of the enzyme ribonuclease.

21

amino acids. They are quite strong bonds and help to hold the tertiary structure of a protein together. In Figure 2.3 on the previous page, the amino acids at positions 26 and 84 are cysteine. The dotted line between them is a disulfide bridge.

When we look at unfamiliar data, it is very important to take the time to think about it carefully and make sure that we understand it. We should be able to answer some basic questions about Figure 2.3.

1 What chemical group is shown by the letter X on the diagram?

*This is an amino (–NH$_2$) group. Look back at Figure 1.14 on page 11. You can see that the dipeptide has a –COOH group at one end and an –NH$_2$ group at the other. The same is true of a polypeptide. There is always a –COOH group at one end and an –NH$_2$ group at the other. Since the –COOH group is shown on the diagram, position X must be where there is an –NH$_2$ group.*

2 Figure 2.3 shows one form of the enzyme ribonuclease. How do you think other forms of ribonuclease will be different?

*Different forms of ribonuclease will have different amino acids in different positions. In other words, each form will have a slightly different primary structure. However, we would not expect them to differ much, because they are all forms of the same enzyme, ribonuclease.*

3 All molecules of this form of ribonuclease have the same tertiary structure. Use information from the diagram to explain why.

*This should be quite simple, if you have understood the diagram. All molecules of this form of ribonuclease will have the same sequence of amino acids. The cysteine molecules will therefore always be in the same positions and the disulfide bridges will form in the same place.*

Hopefully, you have understood what the diagram in Figure 2.3 tells you. We will now look at the steps in Anfinsen's investigation.

- Anfinsen started by measuring the activity of untreated ribonuclease.
- He then treated the ribonuclease with mercaptoethanol. This substance broke the disulfide bridges in the ribonuclease molecules. He measured the activity of the treated ribonuclease.
- Finally, he removed the mercaptoethanol. As a result, the disulfide bonds re-formed. His results are shown in Figure 2.4.

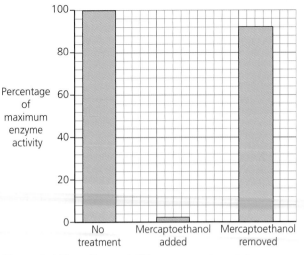

**Figure 2.4** The effects of different experimental treatments on the rate of reaction of ribonuclease.

Again, the data shown in this bar chart are probably unfamiliar. We need to take the time to make sure that we understand the graph. We will start by looking at the axes. The x-axis, the horizontal one, shows the three different treatments. The y-axis, the vertical one, shows the enzyme activity. This is given as the percentage of maximum enzyme activity, so a value of 100 represents the fastest that the enzyme could possibly react.

To make sure that we really understand the information in the graph, we should ask the following, for example.

4 What does the second bar show?

*When the ribonuclease is treated with mercaptoethanol, the enzyme is not very reactive. It shows only about 2% of its maximum activity.*

5 Why did mercaptoethanol have such an effect on the activity of ribonuclease?

*The tertiary structure of ribonuclease is held together by disulfide bridges. These have been broken, so the polypeptide chain loses its shape. We say that it is denatured. This means that the site into which the substrate molecules fit also loses its shape. As a result, the substrate will no longer bind to the enzyme. Not surprisingly, the activity of the enzyme falls.*

6 What happened when the mercaptoethanol was removed from the ribonuclease?

*The bonds re-formed and the tertiary structure of the enzyme was restored. The enzyme was functional again.*

**TIP**
See Chapter 14 to find out how to draw a bar chart.

## Enzymes, substrates and products

If we mix an enzyme solution with biuret reagent, the solution goes violet in colour. This is the test for proteins, and we can use it to show that enzymes are proteins. Now look at Figure 2.5. It shows some of the biochemical reactions that take place in a typical cell.

Each of the 520 dots is a particular substance and the lines connecting these dots are biochemical reactions. What is really important to understand is that each of these reactions is controlled by a different enzyme, so a single cell contains hundreds of different enzymes. These enzymes differ quite a lot in size. Some are rather small molecules and some are relatively enormous, but all of them are proteins.

In addition, each enzyme has a unique shape, or tertiary structure. Somewhere on the surface of the enzyme, a group of amino acids forms a pocket. This pocket is the **active site** of the enzyme. When an enzyme catalyses a particular chemical reaction, a substrate molecule collides with the active site and binds with it to form an unstable intermediate substance called an enzyme–substrate complex. This complex then breaks down to the product molecules. The enzyme molecule is not used up in the reaction. It is now free to bind with another substrate molecule.

An enzyme–substrate complex forms and breaks down rapidly, forming a complex that lowers the activation energy necessary to trigger the reaction. Scientists suggest different ways in which the activation energy might be lowered. The enzyme–substrate complex might bring together substrate molecules in positions that allow the reaction to take place more easily. Or it might put the substrate molecule under stress so that bonds break more readily.

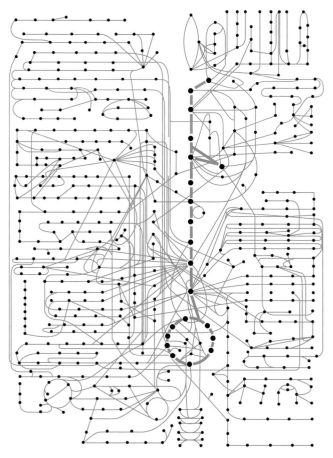

**Figure 2.5** Believe it or not, this is a simplified diagram of some of the biochemical reactions that take place in a single cell. The dots represent different substances and the reactions are shown as lines linking the dots. Each one of the reactions is controlled by a different enzyme.

**Active site** Part of an enzyme molecule, not part of the substrate. The active site is the part of the enzyme molecule into which the substrate fits during a biochemical reaction. Active site and substrate therefore have complementary shapes; they do not have the same shape.

## Enzymes and models

Scientists use models to help them explain their observations. We have known for a long time that enzymes are specific, meaning that each enzyme catalyses just one type of reaction with one type of substrate. Amylase, for example, is an enzyme that hydrolyses starch. It breaks down starch to maltose. If we mix amylase and protein, however, nothing will happen, even though the breakdown of proteins to amino acids also involves hydrolysis. Amylase only hydrolyses its specific substrate, starch. It won't hydrolyse any other substance.

### Induced-fit model

We now know a lot about protein structure. For example, new techniques reveal that proteins are not rigid structures and we have found out that various parts of an enzyme molecule move in response to a change in its environment. Some of these movements are small. Others are quite large and happen when the substrate binds to the enzyme.

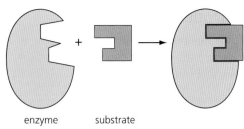

enzyme      substrate

**Figure 2.6** In the induced-fit model, the substrate does not fit precisely until it binds with the active site of the enzyme. The active site then changes shape and moulds round the substrate, allowing a precise fit.

We now have a model to explain how enzymes work, called the induced-fit model, which is shown in Figure 2.6.

In this model, before the substrate binds to the enzyme, the substrate and the active site are not precisely complementary in shape. As the substrate binds, the active site changes shape and moulds closely round the substrate. It is a bit like a sock and a foot. Before you put your sock on, it is not foot-shaped at all. Put it on and it moulds round your foot.

The induced-fit model is now favoured over the lock-and-key model that you learned about at GCSE. In that model of enzyme action, the specific shape of the substrate is said to be a precise fit for the rigid shape of the active site of the enzyme, just like a key in a lock.

## Extension

### Evidence for the induced-fit model

It is one thing to come up with a model to explain how enzymes work, but in order to make real scientific progress we need to find evidence to support it. Is there any evidence for the induced-fit model? We will look at one piece of evidence.

Hexokinase is an enzyme that catalyses the reaction shown below.

glucose + ATP → glucose 6-phosphate + ADP

Hexokinase transfers a phosphate group from a substance called ATP to a glucose molecule. This produces glucose 6-phosphate and ADP. Scientists used a technique called X-ray diffraction, which enabled them to form three-dimensional pictures of molecules. They investigated what happened to the shape of hexokinase when it binds to glucose.

Look at Figure 2.7. It shows you that the active site has changed in shape and has moulded round the glucose molecule.

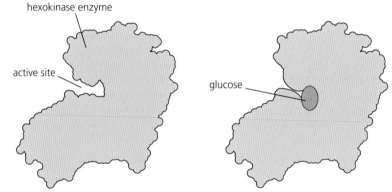

hexokinase enzyme

active site

glucose

**Figure 2.7** The diagram shows the shape of a hexokinase molecule before and after it has bound with its glucose substrate. Look at how the active site of the enzyme has changed shape.

---

**TEST YOURSELF**

1 Name the substrate and product(s) in the following reaction. Lactase catalyses the hydrolysis of lactose to glucose and galactose.
2 Use a diagram to explain why amylase can hydrolyse starch but cannot hydrolyse cellulose.
3 A student wrote 'Maltase forms an enzyme–substrate complex with maltose because maltose is the same shape as the enzyme's active site'. What is wrong with this statement?
4 In the Haber process, nitrogen and hydrogen react together to form ammonia, using a catalyst as well as 15–25 megapascals (Mpa) pressure and 300–550°C. Nitrogen-fixing bacteria have an enzyme called nitrogenase that enables them to carry out the same reaction at atmospheric pressures and normal soil temperature. How does the enzyme do this?

# The properties of enzymes: active sites and enzyme action

Enzyme-controlled reactions depend on substrate molecules fitting into the active site of the enzyme. Environmental factors alter the shape of the active site and therefore alter the rate of a reaction. Such factors include high temperatures, variations in pH and the presence of inhibitors.

## High temperatures

Each enzyme has an optimum temperature. At this temperature, its rate of reaction is around its maximum. If we increase the temperature above its optimum, we increase the kinetic energy of the enzyme molecules. As a result, they vibrate more vigorously, and this breaks the hydrogen and ionic bonds that maintain the tertiary structure of the enzyme molecules. Once these bonds have broken, the enzyme changes its shape. This is called denaturation. Its active site also changes shape and, as a result, substrate molecules will no longer fit and form an enzyme–substrate complex. For example, the saliva of many people contains amylase. This enzyme most actively digests starch at body temperature. When very hot food is eaten, very little will be digested in the mouth as the temperature is too high.

---

**TIP**

Remember that when talking about enzymes you should refer to the 'active site', whereas for other types of protein you should refer to the 'binding site'.

---

**Denaturation** A permanent change to the shape of an enzyme or other protein that occurs as a result of breaking the bonds that maintain its tertiary structure. This results from factors such as high temperature or large changes in pH. Denaturation only refers to changes in tertiary structure, not to the hydrolysis of proteins into the amino acids from which they are made.

---

## EXAMPLE

### Temperature and enzymes

Acetylcholinesterase is an enzyme that exists in several different forms in different animal species. It is found in the nervous system of many animals including fish.

Table 2.1 shows the temperature at which the rate of reaction of this enzyme is at its maximum in three different species of fish.

**Table 2.1** The temperature at which the rate of the reaction catalysed by acetylcholinesterase is at its maximum in three different species of fish.

| Species | Habitat | Temperature at which the rate of reaction is at a maximum/°C |
|---|---|---|
| Icefish | Under the ice in the Antarctic | –2 |
| Trout | Freshwater lakes and streams in Britain | Between 2 and 18 |
| Grey mullet | Mediterranean Sea | 25 |

1 Describe the relationship between the temperature of the fish's habitat and the temperature at which the rate of reaction of acetylcholinesterase is at a maximum.
   *The temperature at which the enzyme's rate of reaction is fastest is about the same temperature that the fish's habitat will be for most of the time.*

2 Trout have two forms of acetylcholinesterase, which are active under different environmental conditions. One has a maximum rate of reaction at 2°C; the other has a maximum rate of reaction at 18°C. Suggest the advantage of this to trout.
   *One form of acetylcholinesterase will be more active in the winter months and the other form will be more active in the summer months when the environmental temperature is higher.*

3 Acetylcholinesterase is found in icefish and grey mullet, but it has slightly different properties in each species of fish.
   a) The acetylcholinesterase in both fish breaks down a substance called acetylcholine. Explain why it is able to do this.
      *The enzyme has an active site that is the right shape for acetylcholine to fit into.*
   b) Use your knowledge of protein structure to explain why the acetylcholinesterase found in these two fish has slightly different properties.
      *Each enzyme may have a slightly different amino acid sequence so that hydrogen bonds and disulfide bridges form in slightly different places, giving a slightly different tertiary structure.*

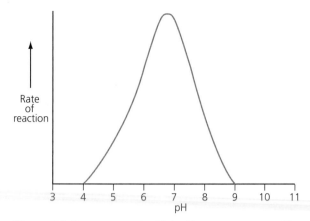

**Figure 2.8** Enzymes work efficiently over a narrow pH range. This graph shows the effect of pH on a typical enzyme from a human cell.

## pH

The pH of a solution is a measure of its hydrogen ion concentration. The higher the concentration of hydrogen ions (H⁺), the lower the pH, and the more acid the solution. pH is a logarithmic scale, so as the pH scale falls one point from 7 to 6, the concentration of hydrogen ions increases 10 times.

Look at the graph in Figure 2.8. It shows that the rate of a reaction rises to a peak at pH 6.5. It then falls sharply. The peak value of 6.5 is the optimum pH for this enzyme.

Changing the pH above or below the optimum of an enzyme affects the rate at which the enzyme works. The change in pH alters the concentration of hydrogen ions (H⁺) or hydroxyl ions (OH⁻) in the surrounding solution. If the change is small, the main effect is to alter charges on the amino acids that make up the active site of the enzyme. As a result, substrate molecules no longer bind. A large change in pH breaks the hydrogen and ionic bonds that maintain the tertiary structure of the enzyme. The result is that the enzyme is denatured.

## Inhibitors

Inhibitors are substances that slow down the rate of enzyme-controlled reactions. Competitive inhibitors are molecules that are very similar in shape to the substrate of the enzyme, as shown in Figure 2.9. They are described as competitive inhibitors because they compete with the substrate for the active site. They fit into the active site and block it. This prevents substrate molecules from entering and stops an enzyme–substrate complex being formed.

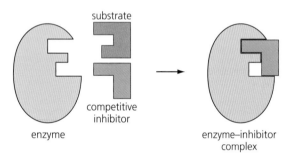

**Figure 2.9** A competitive inhibitor competes with substrate molecules for the active site of the enzyme.

Figure 2.10 shows how non-competitive inhibitors work. Notice that a non-competitive inhibitor doesn't fit into the active site of the enzyme and block it in the way that a competitive inhibitor does. Instead, it binds somewhere else on the enzyme. This causes the enzyme, including the active site, to change shape and, as a result, substrate molecules no longer fit and so no enzyme–substrate complex is formed.

The higher the concentration of a competitive inhibitor, the greater the degree of inhibition. In other words, the effect of a competitive inhibitor can be reduced by adding more substrate. On the other hand, changing the concentration of substrate has no effect on the degree of inhibition shown by a non-competitive inhibitor.

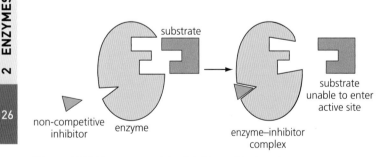

**Figure 2.10** Non-competitive inhibitors bind to the enzyme somewhere other than the active site. This causes the active site to change shape.

**TEST YOURSELF**

**5** Uric acid is a substance that is made in the body. Gout is a painful condition caused when too much uric acid is produced and crystals form in the joints. Gout can be controlled by a drug called allopurinol. Allopurinol is a competitive inhibitor. Suggest how allopurinol controls gout.

**6** Ethylene glycol is a substance found in antifreeze. Sometimes it is consumed accidentally. Ethylene glycol itself is not poisonous but it is broken down in the body by the enzyme alcohol dehydrogenase to substances that are toxic. One way for doctors to treat ethylene glycol poisoning is to give the person a drink containing alcohol. Suggest how this works to prevent the poisoning becoming fatal.

**7** Figure 2.11 shows the effect of a competitive or non-competitive inhibitor on an enzyme-controlled reaction. Explain why adding more substrate reduces the effect of a competitive inhibitor, but adding more substrate does not reduce the effect of a non-competitive inhibitor.

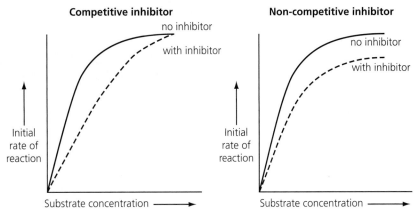

**Figure 2.11** The effect of a competitive or non-competitive inhibitor on an enzyme-controlled reaction.

**8** Copy the graphs shown in Figure 2.11 and add a line to each to show the effect of adding a greater amount of inhibitor.

# The properties of enzymes: collisions come first

An enzyme and its substrate must come together before an enzyme-controlled reaction can take place. They must collide with each other with enough energy to break existing chemical bonds and form new ones. The greater the number of successful collisions in a given period of time, the faster will be the rate of reaction. Increasing the temperature and increasing the concentration of the substrate or of the enzyme can increase the probability that a successful collision will take place and that an enzyme–substrate complex will form.

## Temperature

We have already seen that high temperatures denature enzymes and stop them from working. At temperatures below the optimum, an increase has a different effect. An increase in temperature increases the kinetic energy of the enzyme and substrate molecules. As a result, they move faster. This increases the probability that enzyme and substrate molecules will collide with each other. In most enzyme-controlled reactions, a rise of 10°C more or less doubles the rate of reaction, provided that temperature stays within an acceptable range around the optimum.

**Figure 2.12** The tuatara is a New Zealand reptile. Like other reptiles, the tuatara cannot regulate its body temperature independently of the temperature of its environment (it is an ectotherm). The tuatara is active and its enzymes can digest food at temperatures as low as 6°C. With an increase in temperature, its enzymes work faster and it digests its food faster.

## Extension

In reality, denaturation occurs below the optimum; the optimum is the balance between slowing due to denaturation and increasing due to more collisions.

This is because there are two rates of two reactions: the rate of denaturation and the rate of enzyme–substrate reaction. Both are temperature dependent.

An increase in temperature therefore increases the rate of reaction until the temperature reaches an optimum value, and then the rate of denaturation also increases. Around 10°C above the optimum temperature, although the initial rate of reaction is very fast, it soon falls as the enzyme quickly becomes denatured. This is summarised in Figure 2.13.

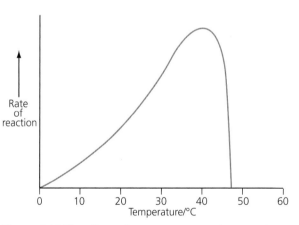

**Figure 2.13** The effect of temperature on the rate of reaction of an enzyme–controlled reaction.

## Rates of reaction

The rate of a reaction is how fast it is going over a given time. Catalase catalyses the reaction in which hydrogen peroxide is broken down to water and oxygen. One way of following the progress of this reaction is to measure the volume of oxygen produced.

## EXAMPLE

### Using catalase to measure rates of reaction

1 Suppose 25 cm³ of oxygen was produced in 20 seconds. What would be the rate of this reaction over this 20 second period?
*The rate would be the volume of oxygen produced divided by the time the reaction took. In other words, 25/20 or 1.2529cm³ per second.*

2 When we carry out this reaction in a tube in a laboratory, its rate changes. As time progresses, it slows down until it comes to a complete stop. Explain why the rate of the reaction slows down. Use your knowledge of the way in which enzymes work.
*The number of hydrogen peroxide molecules will get fewer and fewer as the substrate is broken down. The number of molecules of catalase, however, does not change. This means that there will be fewer collisions between substrate molecules and the active sites of the enzyme molecules. Fewer collisions mean a slower rate of reaction.*

The graph in Figure 2.14 shows a progress curve for a reaction in which catalase is involved in breaking down hydrogen peroxide.

3 Calculate the rate of the reaction between 10 and 20 seconds. Explain how you arrived at your answer.
*Using a graph to calculate the rate of a reaction is really no different from the calculation that you have just carried out. In this case, you read off the total volume of oxygen produced between 10 and 20 seconds. This is 20 cm³ – 10 cm³. 10 cm³ of oxygen is therefore produced in 10 seconds, so the rate is 10/10 or 1 cm³ per second.*

Suppose, however, that you want to look at what is happening a little later on when the rate is changing.

4 Calculate the rate of the reaction at 30 seconds.
*You cannot divide the volume produced by 30. This will give you the mean rate over the whole 40 second period. You want the rate at a particular time, 30 seconds. Figure 2.15 shows what you should do.*

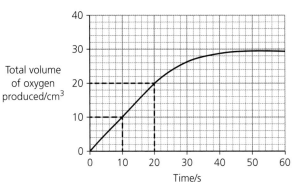

**Figure 2.14** Graph showing the breakdown of hydrogen peroxide by catalase.

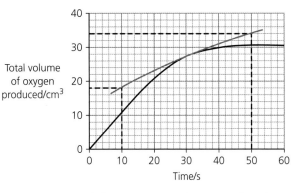

1. Draw a tangent to the curve at the point at which you are interested. Here it is 30 seconds.
2. Use the tangent to read the volume of oxygen produced. Here it is 34 – 18 or 16 cm³.
3. Use the tangent to read off the time in which this volume is produced. Here it is 50 – 10 or 40 s.
4. Calculate the rate by dividing the volume by the time. Here it is 16/40 or 0.4 cm³ s⁻¹.

**Figure 2.15** Advice on answering questions: calculate the rate of the reaction at 30 seconds. Note that a tangent is a straight line that touches but doesn't cross a curved line.

## REQUIRED PRACTICAL 1

### Investigation into the effect of a named variable on the rate of an enzyme-controlled reaction

**Note: This is just one example of how you might tackle this required practical.**

A student decided to investigate the effect of pH on a protein-digesting enzyme. She set up several boiling tubes containing a mixture of $2\,cm^3$ of a protease enzyme and $5\,cm^3$ of a buffer solution as follows. In each tube she placed a glass capillary tube containing solidified egg white (a protein). She measured the length of egg white in each capillary tube before putting it in each tube. She left all the tubes, stoppered, in an incubator at 30°C for 12 hours. After this time, she removed the capillary tubes of egg white from the boiling tubes and measured the length of egg white remaining. The results are shown in Table 2.2.

### SAFETY

Avoid direct contact with the enzymes in general, and particularly proteases. Enzymes can cause allergic reactions, and sometimes sensitisation. Wear eye protection, and wash off splashes from skin. People known to be sensitive to enzymes should avoid the activity, or wear gloves (they should be removed carefully to avoid contamination). Plant or fungal proteins should be used if possible, and care taken when measuring protein cylinders, to avoid dislodging the material.

**Table 2.2**

| pH | Initial length of egg white in tube/mm | Final length of egg white in tube/mm |
|---|---|---|
| 4.8 | 54 | 47 |
| 5.6 | 50 | 45 |
| 5.8 | 52 | 42 |
| 6.2 | 54 | 40 |
| 6.6 | 52 | 32 |
| 6.8 | 53 | 27 |
| 7.2 | 52 | 17 |
| 7.6 | 48 | 7 |
| 8.2 | 47 | 15 |
| 8.6 | 52 | 23 |
| 9.0 | 54 | 37 |
| 9.6 | 53 | 49 |

1 Calculate the percentage of egg white that has been digested in each tube.
2 Plot a graph showing the dependent variable on the *y*-axis against the independent variable on the *x*-axis.
3 Why was it important to calculate the percentage of egg white digested, rather than length of egg white in the tube?
4 Suggest a suitable control for this investigation. Explain why this is needed.
5 Why was it necessary to put stoppers in the tubes before incubating them?
6 The student used a buffer solution to create the correct pH in each tube. Explain why she used a buffer, rather than adding acid or alkali.
7 The student decided that the optimum pH for this enzyme was 7.6. Do you agree with this? Give reasons for your answer.

# Limiting factors and substrate concentration

The shape of the curve in Figure 2.16 is one that you will often come across. It is an example that has nothing to do with biology. Supporters are going to a football match. To get into the ground, they have to pass through turnstiles. The graph shows the rate at which the supporters get into the ground plotted against the number of people outside who are trying to get in.

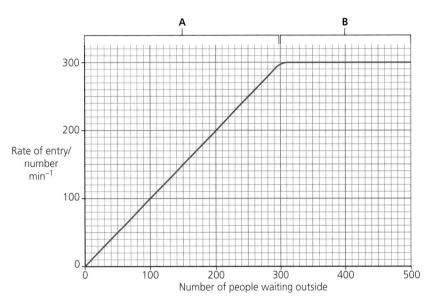

**Figure 2.16** The rate of entry into a football ground plotted against the number of people outside.

We have divided this curve into two parts. Look at part A first. It shows that the rate of entry into the ground is directly proportional to the number of people outside. Three hours before the game starts, very few people are trying to get in. They can go straight to a turnstile and walk through. As more and more people arrive, the rate of entry into the ground increases.

There comes a point, however, when there are so many people outside that all the turnstiles are working as fast as possible and queues start to build up. The rate of entry to the ground cannot get any faster. We are now on the part of the curve labelled B. We say that over part A of the curve, the number of people outside is the limiting factor as it limits the rate of entry to the ground. The curve levels out in part B. It does not matter how much faster supporters arrive at the ground, the rate of entry stays the same. Something else is acting as the limiting factor. It is probably the number of turnstiles.

We will now look at a biological example that is based on the same principles. Figure 2.17 shows what happens to the rate of reaction when the concentration of the substrate is increased but enzyme concentration is kept constant. In this reaction, both the temperature and the pH are at their optimum values.

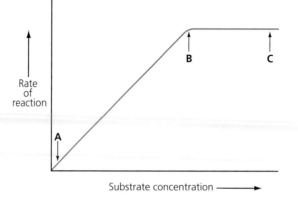

**Figure 2.17** The effect of substrate concentration on the rate of an enzyme-controlled reaction.

---

**The effect of substrate concentration on the rate of an enzyme-controlled reaction**

1 Look at the part of the curve between point A and B on Figure 2.17. What limits the rate of reaction over this part of the curve?
   *The answer is substrate concentration.*

2 What is the evidence from the graph for this answer?
   *As we increase the concentration of the substrate, the rate of reaction also increases.*

3 What causes the rate of reaction to increase over this part of the curve?
   *The more substrate molecules there are, the greater is the probability that one of these molecules will collide* *successfully with the active site of an enzyme and a reaction will take place.*

4 Look at the part of the curve between B and C. How does increasing the substrate concentration after point B affect the rate of reaction?
   *The rate of reaction stays the same.*

5 What caused the rate of reaction to stay the same over this part of the curve?
   *At any one time, all the enzyme active sites are occupied. The enzyme cannot work any faster. The only way that the rate of reaction can be increased is to increase the number of enzyme molecules.*

---

TIP
Look out for curves of the shape shown in Figures 2.16 and 2.17. They are very common in biology and their explanation relies on the same principles every time. We will call what we plot on the *x*-axis *X*, and what we plot on the *y*-axis *Y*.

● The first part of the curve rises. *Y* is limited by *X* because an increase in *X* produces an increase in *Y*.
● The *y*-axis value on the second part of the curve stays constant. On this part of the curve, an increase in *X* has no effect on *Y*. Something other than *X* is limiting *Y*.

## The effect of enzyme concentration on reaction rate

Increasing the enzyme concentration will increase the rate of an enzyme-controlled reaction, provided that there is enough substrate present. This is because adding more enzymes increases the number of active sites for the substrate molecules to collide with. This is shown in Figure 2.18.

Of course, the linear increase in rate of reaction only occurs until substrate concentration becomes limiting.

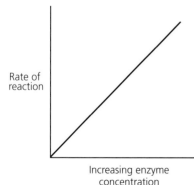

**Figure 2.18** The effect of enzyme concentration on reaction rate.

## TEST YOURSELF

9 As its body temperature increases from 15°C to 30°C, the rate at which a crocodile digests its food increases. What causes this increase?

10 An enzyme works most quickly at a pH of 6.4. A change in pH from 6.4 to 6.8 has a large effect on the rate of the reaction controlled by this enzyme. Use your knowledge of pH to explain why a small change in pH has a large effect on the rate of the reaction.

11 Sketch some axes with the *x*-axis labelled 'enzyme concentration' and the *y*-axis labelled 'rate of reaction' and draw on a line showing the rate of reaction in the presence of excess substrate.

12 Sketch some axes with the *x*-axis labelled 'enzyme concentration' and the *y*-axis labelled 'rate of reaction' and add a line to show the rate of reaction when the substrate concentration becomes a limiting factor.

13 The graph in Figure 2.19 shows the concentration of product when a concentration *A* of an enzyme is added to excess substrate. Sketch a line to show the concentration of product when the concentration of enzyme added is 2*A* but excess substrate is present.

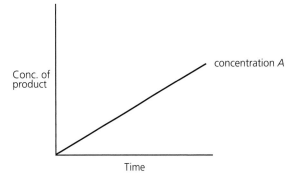

**Figure 2.19** Concentration of product when an enzyme is added to excess substrate.

14 Enzymes used for student practicals are usually stored in the prep room refrigerator. Explain why.

## ACTIVITY

### Making a better oral rehydration solution

This is an investigation that could be carried out in a laboratory. How would you go about it?

Oral rehydration solutions (ORS) can be used by doctors to save the lives of children suffering from diarrhoeal diseases, often following natural disasters such as floods or typhoons. While children are suffering from diarrhoea, they are not absorbing the nutrients they need. As a result, they become weaker and weaker and less able to fight off the effects of the next attack of diarrhoea. Scientists wanted to develop a very effective ORS, one that not only prevents dehydration but also provides the valuable nutrients that a developing child needs.

The scientists investigated an ORS based on the starch in one type of rice flour. They found that it rehydrated diarrhoea sufferers and also helped to overcome malnutrition. However, at high concentrations, a rice-flour ORS is so thick that children cannot swallow it or take it from a feeding bottle. The scientists decided to use the enzyme amylase to digest the starch. This reduces its viscosity (thickness) so that a patient can drink it from a cup or a feeding bottle. They investigated the effect of different concentrations of amylase on reducing the viscosity of rice-flour solution.

## TIP

See Chapter 15, on practical skills, to get ideas about investigating a scientific question.

# Practice questions

1 A student carried out an investigation to find the effect of temperature on the rate of digestion of starch by amylase. She pipetted $5\,cm^3$ of 1% starch solution into each of three test tubes. She then added six drops of iodine solution to each of these tubes. The student took another three tubes. Into each of these tubes she pipetted $2\,cm^3$ of amylase solution and $1\,cm^3$ of a buffer solution at pH 7. She placed one test tube containing starch solution and one test tube containing amylase into each of three water baths, at 10°C, 20°C and 35°C. She left the tubes in the water baths for 10 minutes. After this, she poured the tube containing amylase into the tube containing the starch solution, mixed the contents thoroughly and found the time taken for the blue colour to disappear. The student's results are shown in the table.

| Temperature/°C | Time taken for blue colour to disappear/s |
|---|---|
| 10 | 180 |
| 20 | 125 |
| 35 | 75 |

a) i)  Describe how you could make $20\,cm^3$ of 1% starch solution.  (2)

ii) Explain why the student added a buffer solution to the tubes containing amylase.  (2)

iii) The student left the tubes in the water baths for 10 minutes before mixing them together. Explain why.  (2)

b) Suggest a suitable control for this investigation, explaining why it is necessary.  (2)

c) The student recorded the time taken for all the starch to be digested. Suggest how she could use these results to calculate the rate of reaction. Explain your answer.  (2)

2 The diagram shows a biochemical pathway that takes place inside a cell. It also shows the molecular structures of two of the substances in the pathway, glutamic acid and glucosamine 6-phosphate.

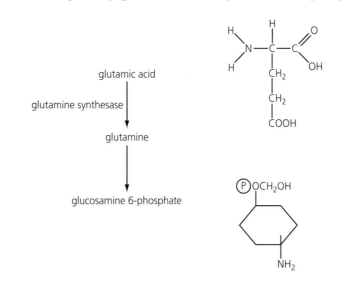

**a)** What type of substance is glutamic acid? Use the diagram to give the reason for your answer. (3)

**b)** Glucosamine 6-phosphate inhibits the enzyme glutamine synthetase. It is a non-competitive inhibitor.

   **i)** What information in the diagram suggests that glucosamine 6-phosphate is not a competitive inhibitor? Explain your answer. (2)

   **ii)** Explain how glucosamine 6-phosphate inhibits glutamate synthetase. (3)

**c)** Suggest a possible advantage to an organism of each of the following.

   **i)** The product formed in a biochemical pathway inhibits one of the enzymes in the pathway. (2)

   **ii)** The enzyme that is inhibited is at the start of the pathway. (2)

   **iii)** The product is usually a non-competitive inhibitor. (2)

**3** Gelatine is a protein. When a gelatine solution cools, it sets to form a jelly. Fresh pineapple juice contains an enzyme that digests protein. A student investigated the effect of pineapple juice on the setting of jelly. He set up three different tubes of gelatine and recorded which had set after 3 hours. The contents of each tube and his results are shown in the table.

| Tube | Contents of tube | Jelly set |
|---|---|---|
| A | 6 cm³ gelatine + 2 cm³ pineapple juice + 2 cm³ water | No |
| B | 6 cm³ gelatine + 2 cm³ pineapple juice + 2 cm³ hydrochloric acid | Yes |
| C | 6 cm³ gelatine + 2 cm³ boiled pineapple juice + 2 cm³ water | Yes |

**a)** Explain why 2 cm³ of water was added to tubes A and C. (2)

**b)** Explain the results of tube A and tube B. (4)

**c)** What was the purpose of tube C? (3)

## Stretch and challenge

**4** How are the properties of enzymes found in organisms that live in extreme environments, such as very cold or very hot places, different from those of enzymes found in most other organisms? How does the molecular structure of enzymes in organisms that live in hot environments allow them to resist denaturation?

**5** How are enzymes used in industry? Focus on the range of uses and also the benefits of using enzymes rather than alternative chemical processes. (For example, in food manufacture; in producing 'stone-washed' jeans; in the production of paper; in cleaning materials, etc.)

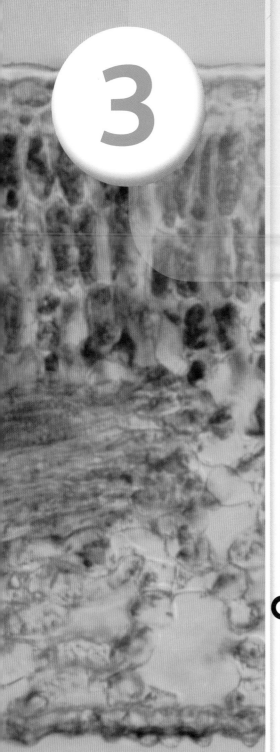

# 3

# Cells

## PRIOR KNOWLEDGE

*Before you start, make sure that you are confident in your knowledge and understanding of the following points.*

- Most human and animal cells have the following parts:
  - a nucleus, which controls the activities of the cell
  - cytoplasm, in which most of the chemical reactions take place
  - a cell membrane, which controls the passage of substances into and out of the cell
  - mitochondria, which are where where respiration occurs
  - ribosomes, which are where protein synthesis occurs.
- Plant and algal cells also have a cell wall made of cellulose, which strengthens the cell. Plant cells often have:
  - chloroplasts, which absorb light to make sugar
  - a permanent vacuole filled with cell sap.
- A bacterial cell consists of cytoplasm and a membrane surrounded by a cell wall; the genes are not in a distinct nucleus.
- Dissolved substances can move into and out of cells by diffusion.
- Diffusion is the spreading of the particles of a gas, or of any substance in solution, resulting in a net movement from a region where they are of a higher concentration to a region with a lower concentration. The greater the difference in concentration, the faster the rate of diffusion.

## TEST YOURSELF ON PRIOR KNOWLEDGE

1 There are more mitochondria in a cell from an insect's wing muscles than in a cell from the lining of the insect's gut. Explain the advantage of this.
2 Give two structural differences between an animal cell and a plant cell.
3 A cell that produces enzymes contains a lot of ribosomes. Explain the advantage of this.
4 What features of cellulose make it useful for building plant cell walls?

## Introduction

It was only when the microscope had been invented that scientists could start to study the detailed structure of living things and see that organisms are made up of cells. The Englishman Robert Hooke (1635–1703) was a brilliant self-taught scientist who made his own microscope. He made a microscope much better than anyone had ever made before, so he could see structures that had never been seen before. You can see his microscope in Figure 3.1.

**Figure 3.1** Robert Hooke's microscope in *Micrographia* (1665).

**Eukaryotic cell** A cell containing a nucleus and other membrane-bound organelles.

**Prokaryotic cell** A cell that does not contain a membrane-bound nucleus or any membrane-bound organelles.

Robert Hooke used his microscope to look at all kinds of things. He made a drawing of cork as seen under his microscope. He noticed that the cork was made up of empty spaces with walls around them. He called them **cells** after the Latin word *cellus*, meaning 'little room'. He calculated the number of cells in a cubic inch to be 1 259 712 000. He did not see all the smaller structures within the cell, because he was looking at the empty spaces between the cell walls in the cork, but he was the first person to realize that cells are the building blocks of living things.

We now know that two of the kinds of cell in living things are eukaryotic cells and prokaryotic cells. Eukaryotic cells are cells that contain a nucleus, such as are found in plants, animals and fungi. Prokaryotic cells include bacterial cells, and these do not contain a nucleus.

## Studying eukaryotic cells

The cells that line the small intestine of a human are called **epithelial** cells. Figure 3.2 shows epithelial cells from the human small intestine. This photograph was taken through an optical microscope.

A

**Figure 3.2** Epithelial cells from the human small intestine. This photograph has been magnified 3300 times.

Look at it carefully and you will see that each cell contains a large structure, which is the **nucleus**. We call animal and plant cells eukaryotic cells because they possess a nucleus. The word eukaryote means 'true nucleus'.

You will also see that the boundary of the cell, where it lines the lumen of the intestine, shows up as a rather fuzzy thick line. If we magnify this a bit more, it does not help a lot. All we get is a slightly thicker fuzzy line. Magnification on its own is not enough. What we need is greater resolution. Magnification is making things larger. Resolution involves distinguishing between objects that are close together. To resolve objects that are close together, we need a microscope that will produce a sharper image.

Light waves limit the resolution of an optical microscope. Using light, it is impossible to resolve two objects that are closer than half the wavelength of the light by which they are viewed. The wavelength of visible light is between 500 and 650 nanometres (nm), so it would be impossible to design an optical microscope using visible light that would distinguish between objects closer than half of this value. That is good enough for a lot of purposes. It is certainly fine for looking at cells from animals and plants. But it won't let you see the very small structures inside a cell. For that we need an electron microscope.

## The transmission electron microscope

If the wavelength of light limits the resolution of an optical microscope, then one solution is to use a beam of electrons instead. Electrons have very much smaller wavelengths than light, so a beam of electrons should be able to resolve two objects that are very close together. That is the way an electron microscope works, as you can see in Figure 3.3.

We cut a very thin section through the tissue that we are going to examine (thin enough to let electrons through). This section, the specimen, is preserved and stained with the salts of heavy metals, such as uranium and lead. Then it is put inside a sealed chamber in the microscope. The air is sucked out of the chamber and this produces a vacuum. Finally, a series of magnetic lenses focuses a beam of electrons through the specimen and produces an image on a screen. Electrons pass more easily through some parts of the section than others. The electrons pass less easily through parts that are stained with heavy metals. This produces contrast between different parts of the specimen.

**Figure 3.3** The diagram shows the main features of a transmission electron microscope.

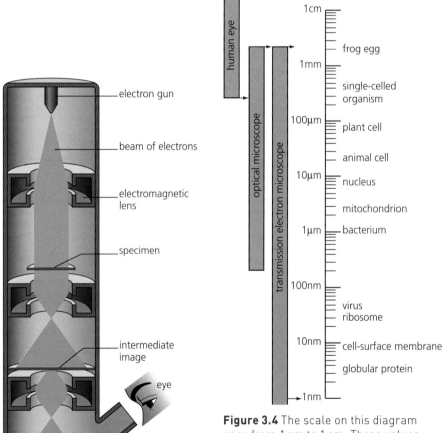

**Figure 3.4** The scale on this diagram goes from 1 nm to 1 cm. These values have been plotted on a log scale because this is the best way of representing such a large range of measurements. The diagram shows that a human eye is able to see large single-celled organisms. With an optical microscope we can see things as small as bacteria. With modern transmission electron microscopes we can see large molecules.

The big advantage of using a transmission electron microscope is its resolving power. It lets us see the structure of a cell in much more detail than we could ever hope to see with an optical microscope. Look at Figure 3.4. It gives you a clear idea of just what can be seen with a human eye, a good quality optical microscope and a transmission electron microscope.

## The scanning electron microscope

A **scanning electron microscope** works in a slightly different way from a transmission electron microscope. In a scanning electron microscope, the electron beam bounces off the surface of the object. It is particularly useful for looking at three-dimensional structures such as viruses.

A high resolving power means that a transmission electron microscope has a useful magnification of up to 100 000 times.

## Electron microscopes have their limitations

You might think that a transmission electron microscope with a magnification of 100 000 times is the perfect instrument to investigate cell structure. However, such a microscope has limitations. Because there is a vacuum inside, all the water must be removed from the specimen. This means that you cannot use a transmission electron microscope to look at living cells. They must be dead. Also, the lengthy treatment required to prepare specimens means that artefacts can be introduced, which look like real structures but are actually the results of preserving and staining.

## ACTIVITY

### Sections through different planes

There are also problems in interpreting what you see. Some of these arise because a very thin slice has to be cut through the specimen. Look at Figure 3.5a. It shows a red blood cell. The shape of a red blood cell is often described as a biconcave disc because it is thinner in the centre and thicker at the edge.

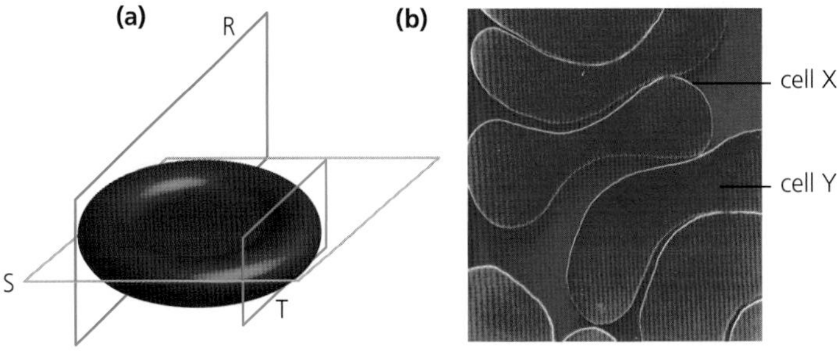

1 Make simple drawings to show what the cut surface of the cell in Figure 3.5a would look like if it were cut through each of the planes shown in the figure.
2 Look at Figure 3.5b. It shows very thin sections of red blood cells. Through which of the three planes shown in diagram (a) is the cell labelled X cut?
3 The cell labelled Y has a bent shape. This bent shape resulted when the cell was sliced. Suggest what caused this bent shape.

**(a)** **(b)**

— cell X

— cell Y

**Figure 3.5** (a) A drawing of a red blood cell. Sections have been cut through it in different planes. (b) A photograph of sections through human red blood cells seen with a transmission electron microscope. A photograph like this is called an electron micrograph.

When the beam of electrons in a transmission electron microscope strikes a specimen, some of the electrons pass straight through and some are scattered by dense parts of the specimen. The parts of the specimen that scatter the electrons appear dark coloured on an electron micrograph.

4 Red blood cells show as a uniform dark colour on an electron micrograph. Explain why.
5 A human red blood cell measures 7 μm in diameter. Calculate the magnification of the electron micrograph in Figure 3.5b (see page 51 for how to do this).

Figure 3.6 shows a scanning electron micrograph of red blood cells.

6 What additional information can you get about red blood cells from Figure 3.6?

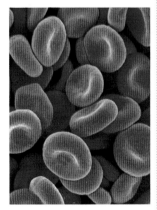

**Figure 3.6** A scanning electron micrograph of red blood cells.

# The ultrastructure of eukaryotic cells

Figure 3.2 shows some epithelial cells from the human small intestine as they appear when looked at with an optical microscope. Now look at Figure 3.7. This shows part of one of the same cells, but it has been taken with a transmission electron microscope, so you can see much more detail.

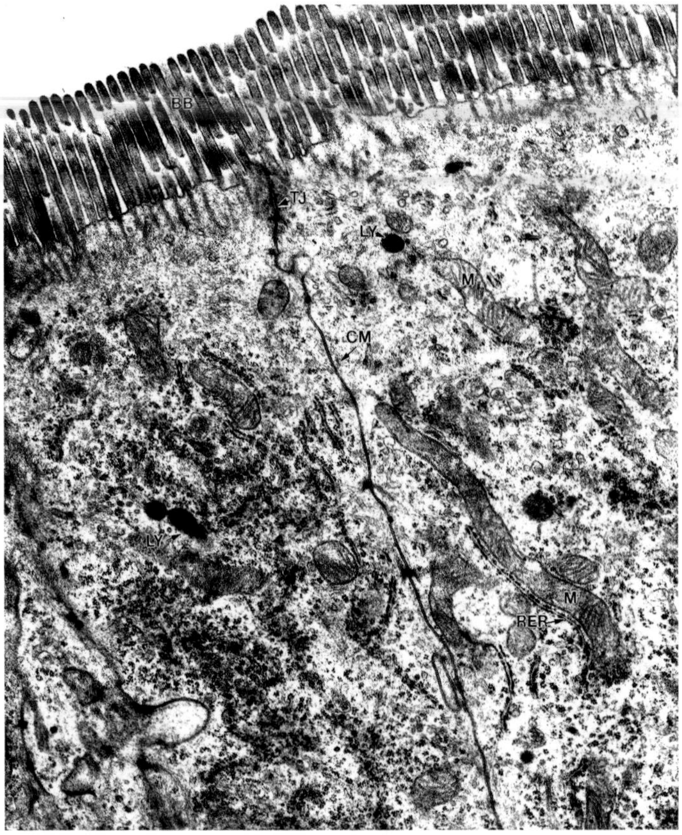

**Figure 3.7** An electron micrograph of an epithelial cell from the small intestine.

Look again at the boundary of the cell, where it lines the lumen of the intestine. It shows that the fuzzy thick line that you saw in Figure 3.2 is made up of tiny finger-like folds in the membrane called **microvilli**. The nucleus can be seen clearly as well. The rest of the cell is made up of cytoplasm in which there are many tiny structures called **organelles**. These organelles have particular functions in the cell. Table 3.1 summarises the functions of the organelles found in an epithelial cell from the small intestine.

Because of the high resolution of a transmission electron microscope, the organelles in the cell can be seen clearly. You could not see most of these organelles with an optical microscope.

## Separating cell organelles

Understanding the structure of an organelle is not the same as understanding its function. To find out about function, biologists need a pure sample containing lots of the organelle that they want to investigate. We separate cell organelles from each other using the process of **cell fractionation**. In this process, a suitable sample of tissue is broken up and then centrifugated at different speeds. Figure 3.8 is a flow chart that summarises the main steps in this process.

The cells in a tissue are broken open in a **homogeniser**. This is a machine rather like a kitchen blender. The tissue is suspended in a buffer solution which keeps the pH constant. This solution is kept cold and has the same water potential as the tissue.

The homogenised mixture is filtered. This removes large pieces of tissue that have not been broken up.

The filtrate is now put in a **centrifuge** and spun at low speed. Large organelles such as nuclei fall to the bottom of the centrifuge tube where they form a **pellet**. They can be resuspended in a fresh solution if they are wanted.

The liquid or **supernatant** is now spun in the centrifuge again. This time, smaller organelles such as mitochondria separate out into a pellet.

**Figure 3.8** The organelles in a sample of tissue can be separated from each other by the process of cell fractionation in a centrifuge. This flow chart shows the main steps in the process.

**Table 3.1** The main organelles found in an epithelial cell from the small intestine.

| Organelle | Main features | Main function |
|---|---|---|
| Cell-surface membrane | The membrane found around the outside of a cell. It is made up of lipids and proteins. | Controls the passage of substances into and out of the cell. |
| Nucleus | The largest organelle in the cell. It is surrounded by a nuclear envelope consisting of two membrane layers. There are many holes in the envelope called nuclear pores. | Contains the DNA, which holds the genetic information necessary for controlling the cell. |
| Mitochondrion | A sausage-shaped organelle. It is surrounded by two membrane layers. The inner one is folded and forms structures called cristae. | Produces ATP during respiration. The molecule ATP is the source of energy for the cell's activities. |
| Lysosome | An organelle containing digestive enzymes called lysozymes. These enzymes are separated from the rest of the cell by the membrane that surrounds the lysosome. | Digests unwanted material in the cell. |
| Ribosome | A very small organelle, not surrounded by a membrane. | Assembles protein molecules from amino acids. |
| Rough endoplasmic reticulum | Endoplasmic reticulum is made of membranes that form a series of tubes in the cytoplasm of the cell. The membranes of rough endoplasmic reticulum are covered with ribosomes. | Synthesises and transports proteins around the cell. |
| Smooth endoplasmic reticulum | Similar to rough endoplasmic reticulum, but the membranes do not have ribosomes. | Synthesises lipids. |
| Golgi apparatus | A stack of flattened sacs, each surrounded by a membrane. Vesicles are continually pinched off from the ends of these sacs. | Packages and processes molecules such as proteins for use in other parts of the cell, or for export to outside the cell. Forms lysosomes. |

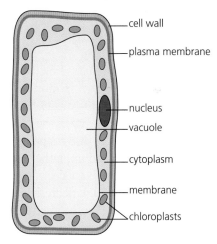

**Figure 3.9** A simple diagram showing a single cell from the palisade layer of a leaf.

## Plant cells

Figure 3.9 shows a single cell from the palisade layer of a leaf. You will notice that, like an animal cell, it is eukaryotic.

Cells in flowering plants have all the organelles that are found in animal cells. However, they have some fundamental differences. A key difference between plants and animals is that plants photosynthesise. This process requires light. To make glucose by photosynthesis, plants have **chlorophyll** to absorb light. Since chlorophyll is only useful in areas exposed to light, it is not present in all plant cells. For example, in underground roots chlorophyll would have no function. In most plants the leaves are where the majority of photosynthesis takes place, in the palisade mesophyll and spongy mesophyll. You can see this in Figure 3.10, overleaf.

**Figure 3.10** A photomicrograph of a vertical section across a leaf, as seen with an optical microscope (x 32).

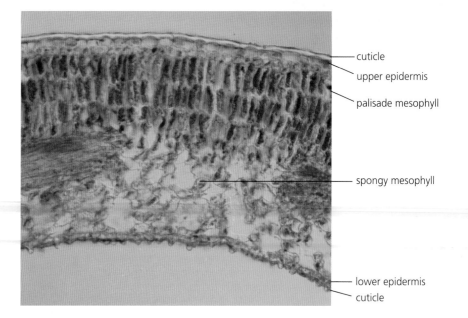

cuticle

upper epidermis

palisade mesophyll

spongy mesophyll

lower epidermis

cuticle

As you can see in Figure 3.11, the chlorophyll is not dispersed throughout the cells but is contained in separate organelles called chloroplasts. When viewed with an electron microscope, a chloroplast can be seen to have a complex structure that adapts it for photosynthesis.

**Figure 3.11** An electron micrograph of a chloroplast (x 16 000).

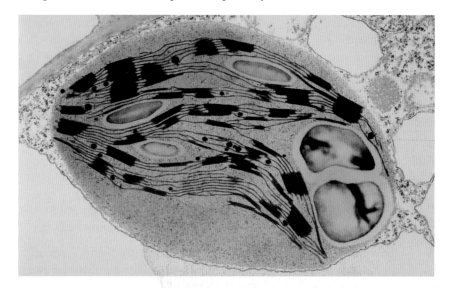

The chlorophyll molecules are embedded in the membranes of disc-shaped structures. In some areas, the disc-shaped structures are stacked up, rather like piles of coins. These stacks show up under low magnification in an electron microscope as dark, grainy patches. There are usually about 50 dark areas in a chloroplast, linked together by a network of the membranes. These structures hold the chlorophyll molecules in positions where the maximum amount of the light that falls on the chloroplast reaches them. The chlorophyll molecules use the light energy to split water into hydrogen ions and oxygen.

The surrounding **stroma** contains enzymes. These enzymes catalyse a series of reactions that use the hydrogen ions, electrons and carbon dioxide to make glucose. The chloroplast is surrounded by a double membrane, which enables control and localisation of substrates and ensures that the enzymes are held inside the chloroplast close to the chlorophyll while allowing free movement of small molecules such as carbon dioxide and water. The structure of the chloroplast ensures that this complex process happens efficiently in a small space with optimum use of available light energy. Excess glucose is converted to insoluble starch and stored temporarily in starch grains in the chloroplast.

## Plant cell walls

You may have noticed that one major difference between animal and plant cells is that an animal cell has no wall, whereas the palisade cell has a wall outside the cell-surface membrane. This wall is fairly thin. It is not rigid like a brick wall, but it does give the cell strength, as it resists being stretched. It is the strength of the cell wall that stops a leaf cell from expanding and taking in so much water that it bursts.

The leaves of most flowering plants stick out from the stems. But, as you know if you have ever forgotten to water houseplants, when a plant is short of water the leaves flop down – they wilt. For the leaves to stay flat, facing the light, they need water that is taken up through the roots. It enters a leaf cell by osmosis and fills the vacuole. The water pushes against the wall and makes the cell firm, just as air pumped into a tyre makes the tyre hard. The cell wall is strong enough to prevent the cell from bursting. Animal cells, however, have no cell wall. If, for example, a red blood cell took in too much water, its cell-surface membrane would burst. Animals have systems that control the water content of the blood and tissue fluid surrounding cells and stop this happening.

All life on Earth exists as cells, which all have common features. This is indirect evidence for evolution. However, not all organisms consist of eukaryotic cells. Prokaryotic cells are described in Chapter 6.

### TEST YOURSELF

**4** Compare the cells shown in figures 3.7 and 3.9. Describe three ways in which the structures of the cells are similar and three ways in which they differ.

**5** Calculate the actual length of the chloroplast in Figure 3.11 in micrometres.

**6** Explain how a rising glucose concentration inside a chloroplast might result in damage to the chloroplast.

### TIP
Cell-surface membranes are found in both prokaryotic and eukaryotic cells. Their structure is the same in both types of cell.

## The cell-surface membrane

You saw in Table 3.1 that the **cell-surface membrane** is made up of lipids and proteins. The commonest lipids found in living organisms are triglycerides. You learned about these in Chapter 1 (pages 7–9). Most of the lipids found in a cell membrane are phospholipids. You will remember that a phospholipid has a 'head' consisting of glycerol and phosphate and a 'tail' containing the long hydrocarbon chains of the fatty acids.

This means that when phospholipids are mixed with water they arrange themselves in a double layer with their hydrophobic tails pointing inwards and their hydrophilic heads pointing outwards. This double layer is called a **phospholipid bilayer** and forms the basis of membranes in and around cells.

### The fluid mosaic model

A cell-surface membrane is only about 7 nm thick, so we cannot see all the details of its structure, even with an electron microscope. Because of this, biologists have produced a model to explain its properties. This is called the **fluid mosaic model**. The model was given this name because it describes how the molecules of the different substances that make up the membrane are arranged in a mosaic. Not all of these molecules stay in one place.

They move around, so we also describe the bilayer as being fluid. All cell membranes have this structure, not just the cell-surface membrane.

**Figure 3.12** A simple diagram showing the structure of a cell-surface membrane. The main components are the phospholipids, which form a bilayer, and proteins.

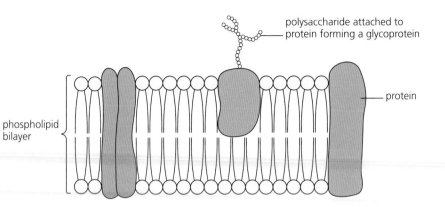

polysaccharide attached to protein forming a glycoprotein

protein

phospholipid bilayer

Look at Figure 3.12. This is a very simple diagram showing a section through a cell-surface membrane. It illustrates how the phospholipids and proteins are arranged. The membrane is based on a phospholipid bilayer. Very small non-polar molecules, and molecules that dissolve in lipids, can pass easily through this bilayer. Water-soluble substances, however, must pass through channels and carriers in the protein molecules that span the membrane. The phospholipid layer therefore forms a very important barrier. Since molecules of some substances are unable to pass through it directly, passage into or out of the cell is controlled by the protein molecules in the membrane.

Some of the proteins move freely in the phospholipid bilayer of the cell-surface membrane. Others are attached to both the cell-surface membrane and structures in the cytoplasm of the cell.

Membrane proteins have a variety of different functions, as follows.

● They may act as enzymes. Enzymes that digest disaccharides are found in the cell-surface membranes of the epithelial cells that line the small intestine.
● They may act as channels through the membrane to allow specific ions or molecules through.
● They act as carrier proteins and play an important part in transporting substances into and out of the cell.
● They act as receptors for hormones. A hormone will only act on a cell that has the right protein receptors in its cell-surface membrane or cytoplasm.
● They act as molecules that are important in cell recognition, and may act as antigens.

Carbohydrates are attached to lipids and proteins on the outside of the cell-surface membrane, forming glycolipids and glycoproteins. They are important in allowing cells to recognise one another. Cholesterol molecules are also found in the cell-surface membrane of animal cells, where they add strength and prevent movement of other molecules in the membrane.

## Diffusion, osmosis and active transport

### Diffusion

Diffusion is the random movement of the ions or atoms or molecules that make up a substance from where they are at a high concentration to where they are at a lower concentration. In other words, particles of the substance diffuse down a **concentration gradient**. Think about what happens if you put a drop of ink into a beaker of water. The ink molecules will gradually spread through the water. They will have diffused from where they were in a high concentration in the original drop to where they are in a lower concentration in the surrounding water.

These particles are moving at random. You can see this quite easily if you look at a tiny drop of toothpaste mixed with water under the microscope. The toothpaste particles move around, twisting and turning. They are moving because the molecules of water in which they are suspended are moving at random and are bumping into them. The kinetic energy that the molecules possess results in this movement.

In a solid, the particles are packed closely together and can only vibrate; in a liquid, the molecules are free to move, but are close together, so bump into each other and change direction; and in a gas, the molecules travel much further before colliding with each other.

Diffusion is also one of the ways in which substances pass into and out of cells. During aerobic respiration, for example, cells produce carbon dioxide. So there is a higher concentration of carbon dioxide inside a cell than outside. Like oxygen, carbon dioxide is a small non-polar molecule and will diffuse from where it is in a high concentration inside a cell through the cell-surface membrane to where it is in a low concentration outside the cell. Surfaces through which diffusion takes place are called **exchange surfaces**.

The **rate of diffusion** is the amount diffused through the surface divided by the time taken. This depends on a number of factors. These include the following.

- Temperature: molecules move faster at higher temperatures, so the higher the temperature, the faster the rate of diffusion.
- Surface area: the greater the surface area of the exchange surface, the faster the total rate of diffusion
- The difference in concentration on either side of the exchange surface. The greater this difference, the faster the rate of diffusion. In the intestine, the blood is continually transporting the products of digestion away from the intestine wall. This ensures a greater concentration gradient and a faster rate of diffusion.
- A thin exchange surface. Diffusion is only efficient over very short distances. Exchange surfaces such as the epithelium of the intestine are just one cell thick.

## TEST YOURSELF

**7** Fish have gills. They use these gills to obtain oxygen from the water by diffusion. Suggest three features of fish gills that are adaptations for efficient diffusion.

**8** Which type of microscope would be best to view:
  **a)** a single-celled organism swimming through some pond water?
  **b)** the surface of a bacterial cell?
  **c)** the internal structure of a chloroplast?

**9** Small, lipid-soluble substances enter cells most quickly. Explain why.

**10** In cell fractionation, why is a buffer solution used?

## Facilitated diffusion

Large, water-soluble molecules such as those of glucose cannot pass directly through the phospholipid bilayer of a cell-surface membrane. They need to be taken across by carrier proteins in the membrane. These carrier proteins have a binding site on their surface, which has a specific shape. A glucose carrier, for example, has a binding site into which only glucose molecules will fit. In addition, different sorts of cells have different carrier proteins. This explains why a particular cell will take up some substances but not others.

Diffusing molecules bind to a carrier protein. The protein changes shape and takes the molecules through the membrane.

Channel proteins help the diffusion of ions. Some ion channels have gates that open and close.

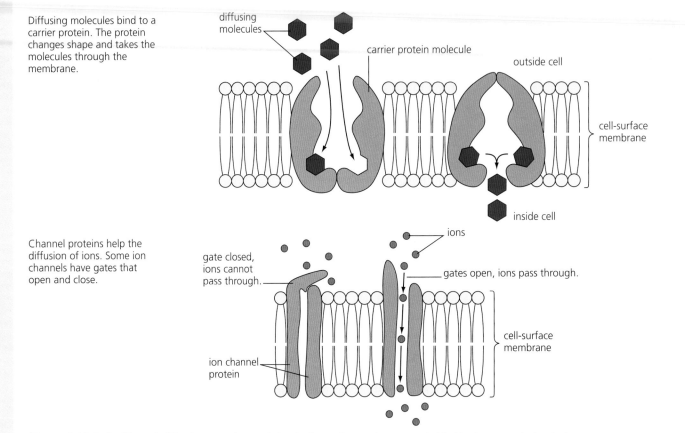

**Figure 3.13** In facilitated diffusion, carrier proteins in the cell membrane assist in the transport of substances into the cell.

Look at Figure 3.13. You can see that carrier proteins change shape when they bind to diffusing molecules, such as glucose molecules. When this happens, the carrier proteins carry the molecules through the membrane and into the cell. This process, in which a protein carrier transfers a molecule that would not otherwise pass through the membrane, is called **facilitated diffusion**. Another kind of facilitated diffusion uses channel proteins. These are proteins that have a water-filled centre which water-soluble molecules and ions can diffuse through. As with simple diffusion, facilitated diffusion relies on the kinetic energy of the molecules. It is also described as a passive process because it does not require hydrolysis of ATP from respiration.

## REQUIRED PRACTICAL 4

### Investigation into the effect of a named variable on the permeability of cell-surface membranes

**Note: This is just one example of how you might tackle this required practical.**

Beetroot cells contain pigments called **betalains** that give the tissue its dark purple-red colour. The pigment is contained in the cell vacuole.

A student decided to investigate the permeability of beetroot membranes. She used a cork borer to cut several 'cores' of beetroot tissue. She cut these cores into slices 2 mm thick and placed them in a beaker of cold water for 2 hours.

The student then took 21 test tubes. She used a graduated pipette to place exactly 5 cm³ of distilled water in each tube. She labelled the tubes with the temperature of the water bath she was going to place them in. She put three tubes in each water bath. The temperatures of the water baths were 0, 25, 40, 50, 65, 85 and 95°C. She left the tubes in the water baths for 5 minutes.

After this, she removed five beetroot discs from the beaker. She handled them carefully using forceps and gently blotted them dry using filter paper. Then she put the discs in the first tube and made a note of the time. She repeated this for each tube.

After exactly 45 minutes, the student shook each tube carefully and gently removed the beetroot discs. She placed a white tile behind each tube and made a note of its colour.

Next, the student set the colorimeter to a blue-green filter (530 nm) and inserted a cuvette containing distilled water. She set the absorbance to 100%.

Then she inserted a sample of the solution from each tube in turn into the colorimeter and measured the percentage absorbance. She obtained the following results.

| Temperature/°C | Colour | Colorimeter reading/% absorption of light | | | |
| --- | --- | --- | --- | --- | --- |
| | | Tube 1 | Tube 2 | Tube 3 | Mean |
| 0 | Clear and colourless | 0.0 | 0.1 | 0.0 | 0.0 |
| 25 | Very pale pink | 3.9 | 4.1 | 4.3 | 4.1 |
| 40 | Very pale pink | 19.3 | 20.4 | 19.9 | 19.9 |
| 50 | Pale pink | 46.2 | 44.3 | 43.5 | 44.7 |
| 65 | Pink | 76.4 | 78.3 | 75.3 | 76.7 |
| 85 | Dark pink | 99.6 | 98.8 | 98.2 | 98.9 |
| 95 | Red | 100.0 | 100.0 | 100.0 | 100.0 |

1 Why were the beetroot discs placed in a beaker of cold distilled water for 2 hours before the investigation?

2 Why did the student handle the discs carefully using forceps, and why were they carefully blotted dry before being added to the tubes of water?

3 Why were the tubes of water left in the water bath for 5 minutes before the discs were added?

4 Why did the student use a red-green filter in the colorimeter?

5 Why did the student put a cuvette containing distilled water into the colorimeter before putting the test solutions in?

6 Plot a graph of these results. Use this graph to describe the effect of temperature on the amount of pigment released from beetroot tissue.

7 Use your knowledge of the structure of cell membranes to explain this effect.

8 Suggest at least two limitations of the student's method. Explain how these might cause small inaccuracies in the results obtained, and suggest how the effects of these limitations could be reduced.

47

## Water potential and osmosis

Look at Figure 3.14, overleaf. It shows water molecules surrounded by a membrane. These water molecules are in constant motion. As they move around randomly, some of them will hit the membrane. The collision of

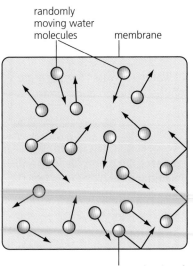

randomly
moving water
molecules

membrane

As molecules of
water move at
random, some
hit the membrane.

**Figure 3.14** Water molecules move at random. Some will hit the surrounding membrane and create a pressure on it. This pressure is the water potential.

**Osmosis** The net movement of water molecules from a solution with a higher water potential to a solution with a lower water potential through a selectively permeable membrane.

the molecules with the membrane creates a pressure on it. This pressure is known as the **water potential** and is measured in units of pressure, usually kilopascals (kPa).

Obviously, the more water molecules that are present and able to move about freely, the greater the water potential. The greatest number of water molecules that it is possible to have in a given volume is in pure water, because nothing else is present. Pure water therefore has the highest water potential. It is given a value of zero. All other solutions will have a value less than this. They will have a negative water potential.

Now look at Figure 3.15. This shows a cell surrounded by distilled water. The cell-surface membrane separates the cytoplasm of the cell from the surrounding water. It is **selectively permeable**. This means that it allows small molecules such as water to pass through but not larger molecules. The cytoplasm of the cell contains many soluble molecules and ions. They attract water molecules, which form a 'shell' round them. These water molecules can no longer move around freely in the cytoplasm. Therefore, there is a much higher concentration of free water molecules in the water surrounding the cell than there is in the cell's cytoplasm.

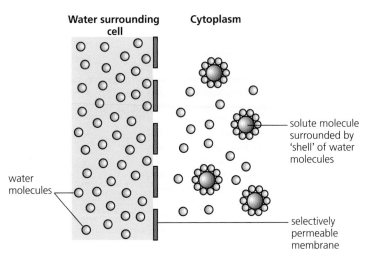

**Water surrounding cell**    **Cytoplasm**

solute molecule
surrounded by
'shell' of water
molecules

water
molecules

selectively
permeable
membrane

**Figure 3.15** In this diagram, the concentration of water molecules that are able to move freely is higher outside the cell than inside the cell. As a result, water will move into the cell by osmosis.

The water potential is higher outside the cell than inside it, and therefore there is net movement of water molecules from the distilled water into the cell. Water molecules will also show net diffusion across a membrane from any solution with a higher water potential to a solution with a lower water potential. This is osmosis. We can, therefore define osmosis in terms of water potential.

If you put an animal cell into a solution with the same water potential as the cell, there will be no net movement of water by osmosis into or out of the cell so it will stay the same.

If you put an animal cell into a solution with a lower water potential than the cell, there will be net movement of water out of the cell and it will shrink in size. The same thing will happen to a plant cell, except that the cell membrane and its contents will shrink away from the cell wall. The gap between the cell membrane and the cell wall will be filled with external solution.

If an animal cell is placed in a solution with a higher water potential than the cell, it will take in water by osmosis and swell up. Eventually it will burst. A plant cell placed in such a solution will also take in water by osmosis, and will swell up to become firm, but it will not burst because the cell wall acts as a protective 'cage' around it, stopping it increasing in volume excessively.

## REQUIRED PRACTICAL 3

## Production of a dilution series of a solute to produce a calibration curve with which to identify the water potential of plant tissue

**Note: This is just one example of how you might tackle this required practical.**

**TIP**
See Figure 1.19 in Chapter 1, page 15, for how to make up a dilution series.

A student was given a 1.0 mol dm$^{-3}$ solution of sodium chloride and a beaker of distilled water. He used this to put 20 cm$^3$ of five different sodium chloride solutions into five Petri dishes as follows: 1.0, 0.75, 0.5, 0.25 and 0.0 mol dm$^{-3}$ sodium chloride.

**1** Copy and complete the table below to show how the student made up these solutions, called a dilution series.

| Dish | Volume of 1.0 mol dm$^{-3}$ sodium chloride solution/cm$^3$ | Volume of distilled water/cm$^3$ |
|---|---|---|
| 1   1.0 mol dm$^{-3}$ sodium chloride | | |
| 2   0.75 mol dm$^{-3}$ sodium chloride | | |
| 3   0.50 mol dm$^{-3}$ sodium chloride | | |
| 4   0.25 mol dm$^{-3}$ sodium chloride | | |
| 5   0.0 mol dm$^{-3}$ sodium chloride | | |

Next, the student used a cork borer to cut several 'cores' of potato. He then cut the cores into discs, with each disc about 2 mm thick. He weighed the discs in batches of five, made a note of the mass, and then placed the discs into one of the dishes. This was repeated for the other four dishes. He left the discs in the dishes for 2 hours.

**SAFETY**
Take care when using knives and cork borers. Your teacher will demonstrate to you the correct, safe technique when cutting cores and discs of potato.

After this time, the student carefully removed the discs from each dish using forceps. He carefully blotted them dry using filter paper, and re-weighed them in the same batches of five. He recorded the results on a table. He calculated the percentage change in mass of the discs.

**2** Copy and complete this table to process the data appropriately.

| Dish | Initial mass/g | Final mass/g | |
|---|---|---|---|
| 1   1.0 mol dm$^{-3}$ sodium chloride | 3.83 | 2.87 | |
| 2   0.75 mol dm$^{-3}$ sodium chloride | 4.07 | 3.62 | |
| 3   0.50 mol dm$^{-3}$ sodium chloride | 4.02 | 2.85 | |
| 4   0.25 mol dm$^{-3}$ sodium chloride | 3.99 | 3.84 | |
| 5   0.0 mol dm$^{-3}$ sodium chloride | 3.96 | 4.73 | |

**3** It was important that the student handled the discs carefully when removing them from the dish and blotting them dry with filter paper. Explain why.

**4** The mass of potato discs placed in each dish was not exactly the same. Does this affect the validity of the student's results? Explain your answer.

**5** Plot a suitable graph of the processed data to find the concentration of sodium chloride solution that has the same water potential as the potato cells.

**6** Explain the changes in mass of the potato tissue. Use the terms 'osmosis' and 'water potential' in your answer.

**7** Identify at least two limitations of this investigation, and suggest how they could be overcome.

## Active transport

Most cells are able to take up substances that are present in lower concentrations outside the cells than inside. Plant cells, for example, contain mineral ions that are present in very small concentrations in the surrounding soil. **Active transport** is a process by which a cell takes up a substance *against* a concentration gradient.

As with facilitated diffusion, protein carrier molecules are involved, and they transport the substance across the membrane. The difference is, however, that active transport requires external energy. This energy comes from molecules of the substance ATP produced during respiration. Cells in which a lot of active transport takes place, such as the epithelial cells lining the small intestine, have large numbers of mitochondria, which produce the necessary ATP.

### TEST YOURSELF

11 Give one difference between simple diffusion and facilitated diffusion.
12 Give one similarity and two differences between facilitated diffusion and active transport.
13 Suggest how channel proteins are specific for certain molecules or ions.
14 List the similarities and differences between simple diffusion and osmosis.
15 Sometimes, in a hospital, a patient who has lost a lot of blood is given saline (sterile solution of sodium chloride) to compensate for the loss of blood. Why is it important that the saline solution have the same water potential as the blood cells?
16 In cell fractionation, why is the buffer used (a) of the same water potential as the cell and (b) ice-cold?

## Prokaryotes and their structure

Bacteria are very small, much smaller than the human cells that some of them infect. Small size is a feature of the cells of all bacteria, and although most features of eukaryotic cells also apply to prokaryotic cells, they do differ from eukaryotic cells in a number of other ways. One is that a bacterial cell does not have a nucleus. It does contain DNA, but this DNA is only present as a circular molecule in the cytoplasm of the cell and not in a nucleus. It is not associated with proteins to form chromosomes.

**Figure 3.16** (a) This photograph shows the main features of a cholera bacterium. (b) The drawing has been made from the photograph to enable you to identify the various features.

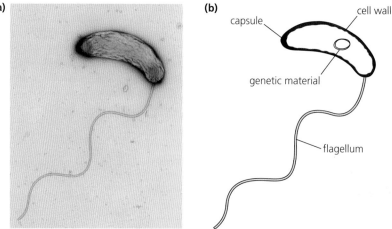

We describe bacterial cells as **prokaryotes** and bacteria as **prokaryotic organisms**. The word prokaryote means 'before the nucleus'. Another feature of bacteria is that some of the DNA they contain, the genetic material, is found in tiny loops called **plasmids**.

A bacterium is surrounded by a cell-surface membrane. (This membrane is also called the plasma membrane.) Outside this membrane is a **cell wall**, but, unlike plant cell walls, it is not made from cellulose. It contains a type of glycoprotein called **murein**. Outside the cell wall there may be another layer. This is the protective **capsule**. This is only present in some bacteria. It helps to stop the bacterium drying out, or from being attacked by white blood cells. It can also store toxins produced by the bacterium. Some bacteria have at least one long whip-like **flagellum**, which helps the cell to move. Flagella are found only in some species of bacteria. Bacteria do not have any membrane-bound organelles, such as endoplasmic reticulum or mitochondria. However, they do have ribosomes that carry out protein synthesis. These ribosomes are smaller than ribosomes in eukaryotes.

Look now at Figure 3.17. This shows a diagram of a bacterium with its main features.

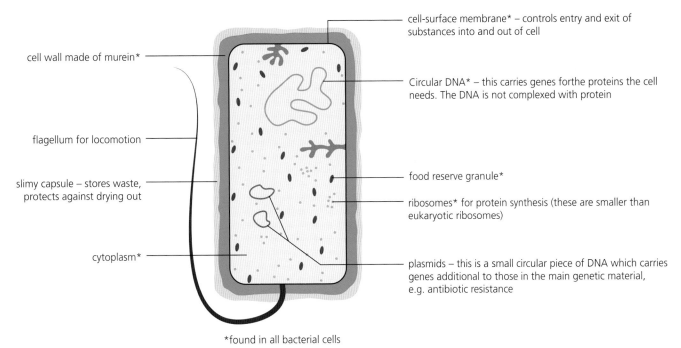

cell-surface membrane* – controls entry and exit of substances into and out of cell

cell wall made of murein*

Circular DNA* – this carries genes forthe proteins the cell needs. The DNA is not complexed with protein

flagellum for locomotion

slimy capsule – stores waste, protects against drying out

food reserve granule*

ribosomes* for protein synthesis (these are smaller than eukaryotic ribosomes)

cytoplasm*

plasmids – this is a small circular piece of DNA which carries genes additional to those in the main genetic material, e.g. antibiotic resistance

*found in all bacterial cells

**Figure 3.17** The main features of a bacterium. Bacteria are tiny, generally ranging in size from 0.1 to 5 μm in length.

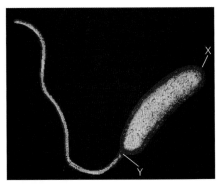

**Figure 3.18** This photograph shows a cholera bacterium (× 10 000).

## How to calculate size from an image

Look at Figure 3.18. It shows a single cholera bacterium magnified 10 000 times. We will use this information to work out its actual size.

The magnification of an object in a photograph is its length in the photograph divided by its real length. We can write this as a simple formula:

$$\text{Magnification} = \frac{\text{size of image}}{\text{size of real object}}$$

If we rearrange this formula, we can calculate the real length of the cholera bacterium in the photograph:

$$\text{Size of real object} = \frac{\text{size of image}}{\text{magnification}}$$

We have been given the magnification, but we need to know the length of the bacterium in the photograph. That is straightforward. All we have to do is use a ruler to measure the length of the bacterium in the photograph. We won't include the flagellum. We will just measure the cell between points **X** and **Y** in the photograph.

It is 30 mm in length, so we substitute these figures in the rearranged formula:

$$\text{Real length} = \frac{30}{10\,000} = 0.003\,\text{mm}$$

The calculation is really very simple, but the units we have used are not very practicable. It is like measuring the cost of a postage stamp in pounds rather than pence. It would be better in this case to give the answer in micrometres (μm). There are a thousand micrometres in a millimetre, so the length of the bacterium is 0.003 × 1000, or 3 μm.

**TIP**

See Chapter 14 to find out how to calculate magnification.

See Chapter 15 to find out how to use measuring instruments such as a graticule and a micrometer.

**TEST YOURSELF**

**17** List the similarities and differences between a prokaryotic cell and a eukaryotic cell.

**18** A bacterial cell is 2 μm long. Its length in a diagram is 40 mm. What is the magnification of the diagram?

# Practice questions

**1** The figure shows a bacterial cell.

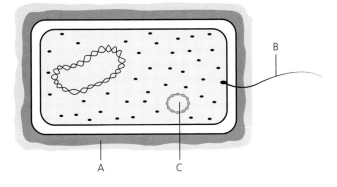

**a)** Name structures A, B and C. (3)

**b)** The following descriptions apply to structures found in a
human cell. Give the name of each structure. (3)

| Name of structure | Description |
|---|---|
| | Synthesises proteins and transports them around the cell. |
| | Contains enzymes that digest substances, such as worn-out organelles or ingested bacterial cells. |
| | Increases the surface area of the cell. |

**2 a)** Copy and complete the table with a tick if the statement
applies or a cross if the statement does not apply. (2)

| | Uses membrane proteins | Requires energy from ATP |
|---|---|---|
| Active transport | | |
| Facilitated diffusion | | |

A student cut a potato tuber in half. She placed the cut part of the potato
downwards in a beaker of distilled water. She then scooped out a hollow in
the top of the potato and placed 10% sucrose solution into this hollow. She
left the potato for 3 hours. The figure shows the results of her investigation.

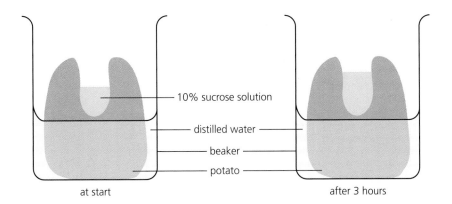

at start                    after 3 hours

**b)** The level of the solution in the hollow at the top of the potato
rose. Use your knowledge of water potential to explain why. (3)

**3** A student cut 30 discs of potato tissue. Each disc was identical in size and shape. She weighed groups of five discs, then put the discs into a dish containing distilled water. This was repeated five more times, with each group of discs being placed in a different concentration of sodium chloride solution. All the dishes were left in a refrigerator for 24 hours. After this, the student carefully dried the potato discs and re-weighed them. She calculated the ratio of the starting mass : final mass for each group of discs. For ease of plotting, a ratio of starting mass to final mass of 0.95 : 1 is plotted as 0.95. The results are shown on the graph.

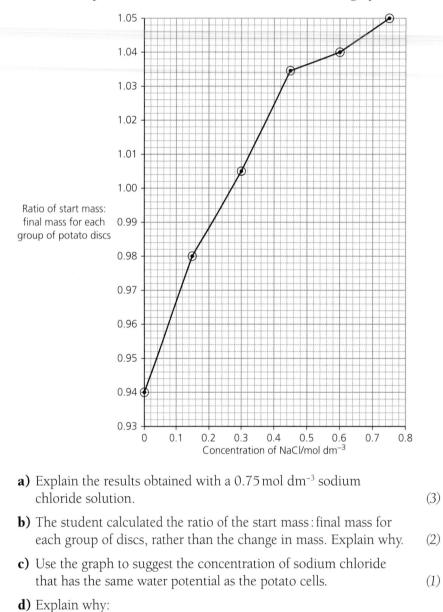

Ratio of start mass: final mass for each group of potato discs

Concentration of NaCl/mol dm$^{-3}$

**a)** Explain the results obtained with a 0.75 mol dm$^{-3}$ sodium chloride solution. (3)

**b)** The student calculated the ratio of the start mass : final mass for each group of discs, rather than the change in mass. Explain why. (2)

**c)** Use the graph to suggest the concentration of sodium chloride that has the same water potential as the potato cells. (1)

**d)** Explain why:

   **i)** the discs were used in groups of five (2)

   **ii)** It was important that the potato discs were identical in size and shape. (2)

**4 a)** The plasma membrane is described as fluid mosaic. Explain why. (2)

The graph shows the rate of diffusion of two different molecules into a cell. (2)

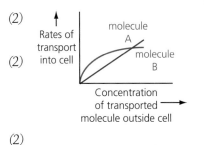

**b) i)** Molecule A enters the cell by simple diffusion, but molecule B enters the cell by facilitated diffusion. Explain the evidence from the graph to show that molecule B enters the cell by facilitated diffusion. (2)

The table gives some information about molecules A and B.

| Molecule | Description |
|---|---|
| | Small, lipid-soluble |
| | Water-soluble ion |

**ii)** Identify molecule A and then copy and complete the table above. Explain your answer. (2)

Beetroot is a vegetable. Its root cells contain a red pigment. In an investigation, discs of tissue were cut from beetroot tissue. All the discs were similar in size and shape. They were rinsed thoroughly, then each disc was placed separately into a test tube containing $10\,cm^3$ of distilled water at various temperatures. After 1 minute the disc was removed from the water and a colorimeter was used to measure the percentage transmission of light through the solution. The lower the transmission, the more red pigment had leaked out of the beetroot disc into the water. The investigation was repeated twice, so that three readings were obtained for each temperature. The results of the investigation are recorded in the table.

| Temperature /°C | Colour of tube after disc removed | Colorimeter reading/% transmission of light | | | |
|---|---|---|---|---|---|
| | | Tube 1 | Tube 2 | Tube 3 | Mean |
| 0 | Clear | 100 | 98 | 99 | 99 |
| 22 | Very pale pink | 94 | 95 | 96 | 95 |
| 42 | Very pale pink | 80 | 77 | 79 | 78 |
| 63 | Pink | 27 | 29 | 31 | 29 |
| 87 | Dark pink | 1 | 1 | 1 | 1 |
| 93 | Red | 0 | 0 | 0 | 0 |

**c) i)** Describe these results. (2)

**ii)** Use your knowledge of the structure of cell membranes to explain these results. (2)

## Stretch and challenge

**5** Describe the detailed structure of chloroplasts and mitochondria. Evaluate the hypothesis that chloroplasts and mitochondria have developed from prokaryotic cells that were engulfed by primitive cells many millions of years ago.

**6** Describe the structure of cells of the archaebacteria. Compare and contrast them with the 'standard' prokaryotic cells of the eubacteria.

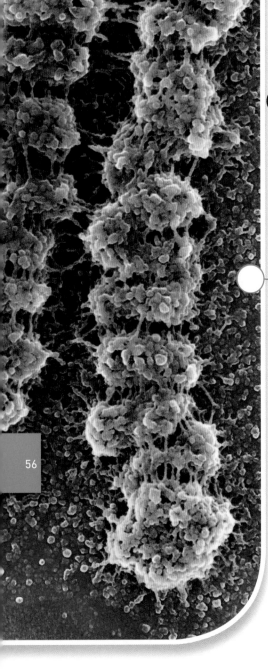

# 4 DNA and protein synthesis

**TEST YOURSELF ON PRIOR KNOWLEDGE**

1 Put the following in order of size, starting with the smallest: gene, nucleus, chromosome.
2 Sketch the shape of a DNA molecule.
3 Mature human red blood cells have no nucleus. Explain the advantage of this.
4 Immature red blood cells are called reticulocytes, and these cells do have a nucleus. Explain why.

## Introduction

The two men in Figure 4.1 are James Watson and Francis Crick. They worked out the structure of **DNA (deoxyribonucleic acid)**. The photo shows them standing by the model of DNA that they produced in 1953.

In 1952, James Watson had completed his PhD in the USA and was carrying out research at the University of Cambridge in the UK. Here he met Francis Crick, who had yet to finish the research for his PhD in physics and was becoming quite bored with it. The two men were fascinated by DNA and spent many hours discussing its possible structure.

**Figure 4.1** James Watson and Francis Crick first proposed the structure of DNA that you will learn about in this chapter. Here, the two researchers are posing in front of the DNA model they built in 1953. Crick is pointing at the model.

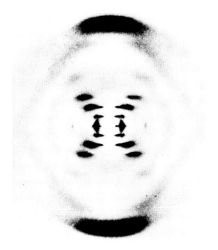

**Figure 4.2** Rosalind Franklin produced this photograph from X-rays diffracted through a DNA molecule on to a photographic plate, and it told her that DNA is a double helix. Unfortunately for her, Watson and Crick saw her evidence and used it before she did.

## What other scientists had found out

They knew from experiments performed in the previous 20 years that DNA is the hereditary material of organisms: a substance that carries information for characteristics from one generation to the next. They knew that DNA was one type of the substances found in cells called **nucleic acids**. They also knew that it contained chemical groups of atoms called deoxyribose and phosphate, and four types of a group called organic bases. These bases are adenine, thymine, cytosine and guanine. Watson and Crick also knew Chargaff's rule: that the number of adenine bases in a molecule of DNA is always the same as the number of thymine bases, and similarly for the pair of bases cytosine and guanine. Maurice Wilkins and Rosalind Franklin working at Kings College in London had evidence that DNA was a helical structure (Figure 4.2).

## What Watson and Crick did

Using solely this information, molecular models and some very inspired guesswork, Watson and Crick came up with the structure of DNA. They published a 900-word report of their proposed structure in the journal *Nature* in April 1953 and won a Nobel Prize in Physiology or Medicine in 1962 for their discovery. Reflecting their reliance on the work of others, their Nobel Prize was shared with Wilkins. Unfortunately, by that time, Rosalind Franklin had died of cancer; otherwise she would also have shared the Nobel Prize.

# DNA structure

## Nucleotides

**Nucleotides** The subunits from which DNA is made. They are formed from a pentose sugar, a phosphate group and a nitrogen-containing organic base.

The structure of DNA and the way that it carries information is the same in all organisms, which is indirect evidence for evolution. Nucleic acids are polymers made up of repeated subunits. In this case, the subunit is called a nucleotide. Figure 4.3 shows the structure of a single nucleotide from DNA. It has three components, as follows.

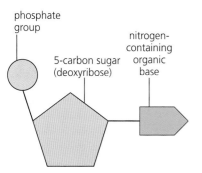

- First there is a **pentose** (five-carbon sugar): DNA gets part of its name from the sugar in each of its nucleotides. The sugar is called **deoxyribose** (which is the sugar ribose with one oxygen atom missing).
- Second is a **phosphate group**. This has a negative charge, which makes DNA a highly charged molecule. This negative charge enables us to separate fragments of DNA by a technique called electrophoresis.
- Third is a **nitrogen-containing organic base**. Each nucleotide contains one of four bases. Two are **adenine** (A) and **guanine** (G) (bases known as purines). The other two are **thymine** (T) and **cytosine** (C) (bases known as pyrimidines). Each base contains some nitrogen atoms in its structure.

**Figure 4.3** A single DNA nucleotide is made from a molecule of a five-carbon sugar (a pentose) called deoxyribose, a phosphate group and an organic base. Nucleotides are the monomers from which nucleic acids are made.

> **TIP**
> You will learn about the technique of electrophoresis during your second year course (see *AQA A-level Biology Year 2 Student's Book*).

## Polynucleotide strands

Figure 4.4 shows how two nucleotides join together by a condensation reaction (see Chapter 1, page 2). You can see that two deoxyribose groups (also called residues) become linked together through one of the phosphate groups. This is called a **phosphodiester bond**. The carbon atoms in a pentose are numbered in a clockwise direction from the one that carries the base. When many nucleotides become linked together like this, they form a polynucleotide. You can see the diagram of a single **polynucleotide** strand in Figure 4.5a. Notice how the pentoses and phosphates form a sugar–phosphate backbone. As we will be more interested in the organic bases, we can simplify the polynucleotide strand to the structure shown in Figure 4.5b.

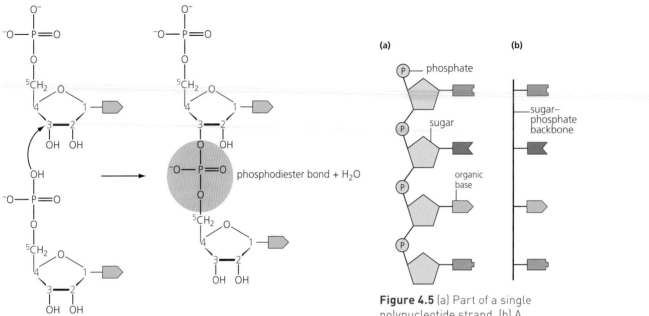

**Figure 4.4** Two nucleotides react to form a dinucleotide. Numbers show the positions of the carbon atoms in the pentoses.

**Figure 4.5** (a) Part of a single polynucleotide strand. (b) A simpler way to represent the same polynucleotide strand.

Hydrogen bond A chemical bond important in the three-dimensional structure of biological molecules. Hydrogen bonds require relatively little energy to break.

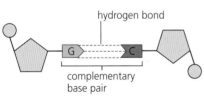

**Figure 4.6** Hydrogen bonds form between the complementary bases of two nucleotides, producing a complementary base pair.

**TIP**

If you are asked to draw a box around the phosphodiester bond in a diagram, include the phosphate atom and all four oxygen atoms of the phosphate group.

## Base pairing

We saw in Figure 4.4 how two nucleotides condense to form a dinucleotide joined by a phosphodiester bond. Figure 4.6 shows a different way that two nucleotides can join together. The two nucleotides with complementary bases (see Table 4.1) are joined by different chemical bonds called hydrogen bonds between the bases. In this way the bases become a complementary base pair. Individually, hydrogen bonds are weaker than the phosphodiester bonds holding the sugar–phosphate backbone together, but collectively they are very strong, which gives stability.

Base pairing occurs only between complementary bases. In DNA, adenine always pairs with thymine, and cytosine always pairs with guanine.

**TIP**

You need to remember that A pairs with T, and C pairs with G. A good way to remember this is to think of initials of people you know, or celebrities you admire, such as footballers. Think of people whose initials are AT or TA, and CG or GC.

## DNA has two polynucleotide strands

Some types of nucleic acid molecule are made of a single polynucleotide strand, like the one shown in Figure 4.5. DNA is not single-stranded; it is made from two polynucleotide strands. The two strands are held together by hydrogen bonds between the complementary base pairs shown in Table 4.1.

**Table 4.1** Complementary base pairs. A and T are complementary bases (A=T pair); C and G are complementary bases (C≡G pair).

| Purine base | | Pyrimidine base |
| --- | --- | --- |
| Adenine (A) | pairs with... | Thymine (T) |
| Guanine (G) | pairs with... | Cytosine (C) |

Look at Figure 4.7, which shows part of a molecule of DNA. You should be able to identify individual nucleotides, the phosphodiester bonds holding together the nucleotides in one polynucleotide strand and hydrogen bonds between complementary base pairs that are holding together the two polynucleotide strands. Look closely at the base sequence of each polynucleotide strand. Notice that one strand has a base sequence that is complementary to the base sequence on the other strand, and that the two strands run in opposite directions. For this reason, we call them **anti-parallel strands**.

Figure 4.8a shows a simpler version of part of a DNA molecule. Here, the sugar–phosphate backbones are shown as single lines. Figure 4.8b shows the final complication of a DNA molecule. The two polynucleotide strands are twisted into a coil called a helix. This diagram shows why a DNA molecule is often referred to as a **double helix**.

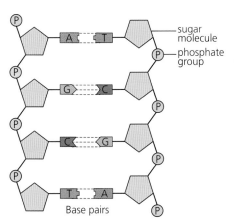

**Figure 4.7** Part of a molecule of DNA showing that it has two anti-parallel strands.

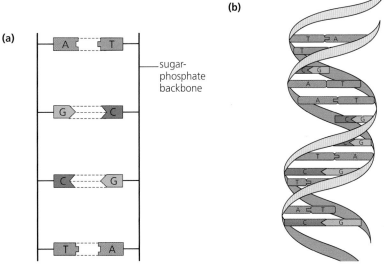

**Figure 4.8** (a) A simpler version of Figure 4.7. (b) A molecule of DNA, showing the double helix of polynucleotide strands.

## DNA is a stable molecule

Since DNA carries the genetic information that cells use to produce their polypeptides, it is important that a DNA molecule does not change. Two types of chemical bond hold DNA molecules together. The first is the phosphodiester bond that joins the phosphate group of one nucleotide to the sugar of the next. Look again at Figure 4.4 to see this. This is a fairly strong (covalent) bond and is not easily broken. The second is the hydrogen bond between bases in a base pair, as shown in Figure 4.8a. Although hydrogen bonds are relatively weak, a single molecule of DNA might be several thousand nucleotides long. Thousands of hydrogen bonds ensure that the two polynucleotide strands are held firmly together.

**TEST YOURSELF**

**1** How many nucleotides are shown in Figure 4.7?
**2** What makes one DNA nucleotide different from another?
**3** A hydrogen bond is a relatively weak bond. Explain why the hydrogen bonds in DNA hold the two strands together relatively strongly.
**4** What is a condensation reaction and what type of condensation bond is formed between nucleotides?
**5** Because DNA forms a helix shape, this means the molecule is coiled up. Explain the advantage of this.
**6** In part of a DNA molecule, 28% of the bases were cytosine. What were the percentages of the other bases?

Scientists were not easily convinced that DNA carries the genetic code. It might be a surprise to hear that, for the first half of the twentieth century, most scientists believed that proteins carried the genetic information. They thought the components of DNA seemed much too simple, as they involved only four bases! In the box below you can see some experiments that changed their minds.

## Extension

### Experiments showing that DNA is the genetic material

In the first experiment, a bacterium called *Streptococcus pneumoniae* was used. It causes pneumonia in humans and other mammals. The bacterium is rod-shaped and has two strains. On agar plates, colonies of the S strain bacterial cells produce an outer polysaccharide coat and appear smooth. Colonies of the mutant R strain bacterial cells lack the polysaccharide coat and appear rough.

A team of scientists injected mice with different combinations of these two strains of the bacterium.

Figure 4.9 shows that mice died from pneumonia when injected with cells of the S strain.

Heat-killed S strain on its own did not cause mice to die, yet mixed with R strain it did. The first team of scientists concluded that the genetic information of the heat-killed S strain was able to get into the live cells of the R strain and transform them into S-type cells. They did not know the chemical nature of this transforming agent.

| live cells of S strain | live cells of R strain | heat-killed cells of S strain | heat-killed cells of S strain and live cells of R strain |
| --- | --- | --- | --- |
| mouse contracts pneumonia | mouse remains healthy | mouse remains healthy | mouse contracts pneumonia |
| colonies of S strain isolated from tissue of dead mouse | colonies of R strain isolated from tissue of healthy mouse | no colonies isolated from tissue of healthy mouse | colonies of R strain and S strain from tissue of dead mouse |

**Figure 4.9** The effect of injecting mice with different combinations of the S strain and R strain of *Streptococcus pneumoniae*.

Some time later, another team of scientists set up an experiment to try to find the nature of the transforming agent. They treated heat-killed samples of the S strain of *S. pneumoniae* with different enzymes. Each enzyme broke down specific molecules within the bacteria. The scientists then mixed each of the extracts from these S strain cells with a different culture of the R strain, and looked at the type of colony that grew on an agar plate. Table 4.2 shows their results.

**Table 4.2** The effect of incubation with different enzymes on the ability of the S strain of *S. pneumoniae* to transform the R strain.

| Experiment | Enzyme used to treat heat-killed cells of S strain of *S. pneumoniae* | Appearance of R strain colonies growing on agar plates |
|---|---|---|
| 1 | Protease | Smooth |
| 2 | Ribonuclease | Smooth |
| 3 | Deoxyribonuclease | Rough |

This research team suspected that three types of molecule found in cells might be the transforming agent: DNA, RNA or protein. Table 4.2 shows that only hydrolysis of the DNA in the S strain extract prevented the R strain being smooth so only DNA could pass on the information needed for the R strain to produce a polysaccharide coat.

Most scientists remained unconvinced, and these results were largely ignored for many years. The results were criticised for several reasons. These included:

- some contaminating protein could have been left in the protease preparation
- DNA might be only part of a pathway that proteins used for transformation.

It took a Nobel Prize-winning experiment to convince scientists around the world that DNA did carry the genetic code. This experiment used a **bacteriophage**. This is a virus that infects and kills bacteria. The virus was the T2 bacteriophage. This bacteriophage infects *Escherichia coli*, a bacterium that commonly grows in the gut of humans. Figure 4.10 shows a single T2 bacteriophage. Notice its simple structure: it has an outer protein capsule surrounding a molecule of DNA. When a T2 bacteriophage infects an *E. coli* bacterium, it multiplies to produce large numbers of bacteriophages that burst the bacterial cell and are released.

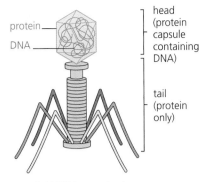

**Figure 4.10** T2 bacteriophages are viruses that infect the bacterium *Escherichia coli*.

The proteins contain sulfur but DNA does not. DNA contains phosphorus, but proteins do not. Each of these elements has a radioactive isotope, $^{32}P$ and $^{35}S$. The team of scientists grew some T2 bacteriophage in the presence of $^{32}P$, which labels their DNA, and some in the presence of $^{35}S$, which labels their proteins. After infecting a culture of *E. coli* with these labelled bacteriophages, the team put samples of the culture in a blender. This removed the bacteriophages from the surface of the bacteria. They then looked to see where the radioactive elements were found. Figure 4.11 shows their results.

Most of the protein was found outside the infected cells and most of the DNA was inside, so it was the DNA that was being used by the infected bacteria to build new bacteriophages. The team went on to show that new bacteriophages released by bursting *E. coli* cells were labelled only with $^{32}P$. This finally convinced scientists that DNA did carry the genetic code.

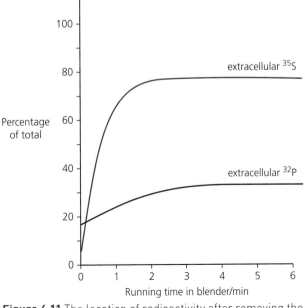

**Figure 4.11** The location of radioactivity after removing the T2 bacteriophage from the surface of infected *E. coli* cells.

# DNA, chromosomes and genes

The DNA in prokaryotic cells is different from the DNA in eukaryotic cells (see Chapter 3, page 50). Prokaryotes contain a single, circular DNA molecule. In addition, they usually contain one or more much smaller circular DNA molecules called plasmids (see Chapter 3, page 51).

In contrast, a single eukaryotic cell always has many, different, molecules of DNA. In eukaryotic cells each DNA molecule is linear and wrapped around proteins called **histones**, and forms a rod-like structure (with two ends) called a **chromosome**. Figure 4.12 shows a chromosome from a eukaryotic cell. In it, the DNA helix is tightly coiled, so that the chromosome looks much thicker than a DNA molecule. Look closely at Figure 4.12. Can you see some of the polynucleotide strands within the thick chromosome?

Each chromosome carries the genetic information for a large number of polypeptides, as well as the information for building the functional RNA molecules needed for protein synthesis (see page 64). The base sequence of DNA coding for a single polypeptide or a functional RNA is called a gene. Look at Figure 4.13. It shows a chromosome as a long string of genes. Each gene has a specific position on a chromosome, called its **locus**. For example, the human gene that codes for pancreatic amylase (see Chapter 8, page 141) is located at its locus on the short arm of chromosome 1. The sequence of bases in many genes encodes the amino acid sequence of a single polypeptide molecule.

It might surprise you to learn that much of the DNA in eukaryotic cells does not code for polypeptides. In fact, less than 2% of human DNA is thought to code for them. Figure 4.14 shows how some apparently non-coding DNA is found within a gene, and some is found between genes. The non-coding regions of DNA within a gene are called **introns**; the coding regions are called **exons**. In between genes, non-coding DNA often contains the same bases sequences repeated many times called non-coding multiple repeats.

**Figure 4.12** An electron micrograph of a chromosome. If you look closely, you can see some bits of the polynucleotide strand that is tightly coiled in the chromosome.

......................................................

**Gene** A base sequence of DNA that codes for the amino acid sequence of a polypeptide or a functional RNA molecule.

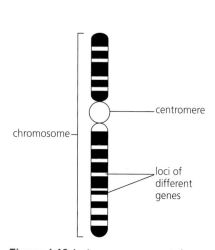

**Figure 4.13** A chromosome contains many genes. Each gene occurs on one chromosome only and occupies a fixed position called its locus.

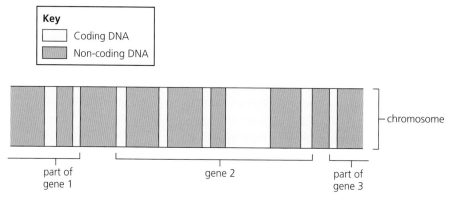

**Figure 4.14** Non-coding DNA is found within a gene and between adjacent genes.

Differences between the DNA found in prokaryotic and eukaryotic cells have been described above. These differences are summarised in Table 4.3. Because of these differences, we should not refer to the genetic material of a prokaryotic cell as a chromosome. However, as a shorthand description, some people refer to 'bacterial chromosomes'.

**Table 4.3** A comparison of the DNA molecules found in prokaryotic and eukaryotic cells.

| Feature of DNA | Prokaryotic cells | Eukaryotic cells |
| --- | --- | --- |
| Relative length of molecule | Short, i.e. few genes | Long, i.e. many genes |
| Shape of molecule | Circular, forming a closed loop | Linear, forming part of a chromosome |
| Number of different molecules per cell | One | More than one |
| Association with proteins | Not associated with proteins | Associated with proteins, called histones |
| Non-coding DNA | Absent | Present within genes (introns) and as non-coding multiple repeats between genes |

**TEST YOURSELF**

**7** Give three ways in which the DNA of prokaryotic cells is different from DNA in eukaryotic cells.

**8** The DNA in a T2 bacteriophage is not associated with histone proteins. Explain why this is important in the investigation shown in Figure 4.11.

**9** Explain the difference between:
  **a)** introns and exons
  **b)** a gene and a locus.

**10** Describe the two types of non-coding DNA found in eukaryotic cells.

# DNA and protein synthesis

**Genome** The complete set of genes in a cell.

**Proteome** The full range of proteins that a cell is able to produce.

An organism's genes are made of DNA. Genes contain the information to synthesise proteins. All of an organism's genes, known as its genome, are contained in every one of its cells. The cell uses some, but not all, of these genes to make a set of proteins. Which proteins it actually makes depends on the type of cell it is. The particular range of proteins that a cell produces using its DNA is known as its proteome. However, before you can understand how genes work, you need to learn about another type of nucleic acid, called **ribonucleic acid** (or **RNA**).

Like DNA, a molecule of RNA is a polynucleotide chain. Figure 4.15 shows a DNA nucleotide, which you are already familiar with, and an RNA nucleotide. Can you spot the difference? Look at carbon atom 2 of the five-carbon sugar. In the DNA nucleotide, it lacks an oxygen atom that is present in the RNA nucleotide. There is another difference that is not shown in Figure 4.15. An RNA nucleotide never has thymine as its base; instead it has uracil.

Figure 4.16 (overleaf) shows the structure of DNA and of two different types of RNA: messenger RNA (mRNA) and transfer RNA (tRNA). How many differences in the structures and compositions of DNA, mRNA and tRNA can you spot in Figures 4.15 and 4.16? Table 4.4 summarises these differences for you.

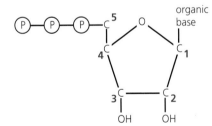

ribonucleotide

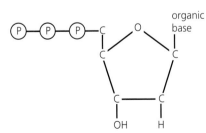

deoxyribonucleotide

**Figure 4.15** The molecular structures of an RNA nucleotide and a DNA nucleotide. RNA contains the sugar ribose; DNA contains the sugar deoxyribose (which has one less oxygen than ribose).

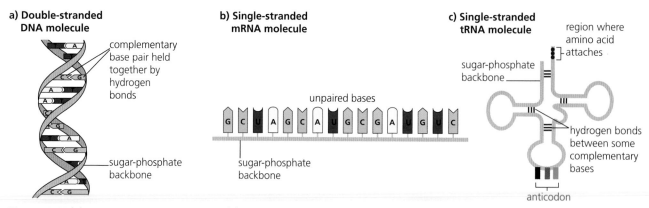

**a) Double-stranded DNA molecule**

complementary base pair held together by hydrogen bonds

sugar-phosphate backbone

**b) Single-stranded mRNA molecule**

unpaired bases

G C U A G C A U G C G A U G U C

sugar-phosphate backbone

**c) Single-stranded tRNA molecule**

region where amino acid attaches

sugar-phosphate backbone

hydrogen bonds between some complementary bases

anticodon

**Figure 4.16** (a) Part of a molecule of DNA, (b) part of a molecule of mRNA and (c) a molecule of tRNA. The polynucleotide chains are represented as single lines. The parts are not to the same scale.

There is a third type of RNA called ribosomal RNA (rRNA) that forms part of the structure of ribosomes.

**Table 4.4** A comparison of DNA, mRNA and tRNA.

| Feature | DNA | mRNA | tRNA |
|---|---|---|---|
| **Nucleotide structure** | | | |
| Pentose sugar | Deoxyribose | Ribose | Ribose |
| Purine base | Adenine and thymine | Adenine and uracil | Adenine or uracil |
| Pyrimidine base | Cytosine or guanine | Cytosine or guanine | Cytosine or guanine |
| **Polynucleotide chain** | | | |
| Number of polynucleotide strands | 2 | 1 | 1 |
| Number of nucleotides in chain | Many millions | Several hundred or thousands | About 75 |
| Hydrogen bonding between complementary base pairs | Present: holds two anti-parallel strands together (A–T, C–G) | Absent | Present in parts of molecule, giving the molecule a clover-leaf shape (A–U, C–G) |

# Adenosine triphosphate

You will have seen **ATP** (adenosine triphosphate) mentioned several times already. For example, mitochondria make ATP during aerobic respiration in eukaryotic cells (see Chapter 3, page 41). ATP is used by cells whenever a process requires energy. Protein synthesis is an example of an energy-requiring process.

ATP molecules have a structure closely related to the RNA nucleotide containing adenine. Figure 4.17a shows how an ATP molecule has the five-carbon sugar ribose and the base adenine, but has three phosphate groups rather than just one. For this reason, it is called a phosphorylated nucleotide.

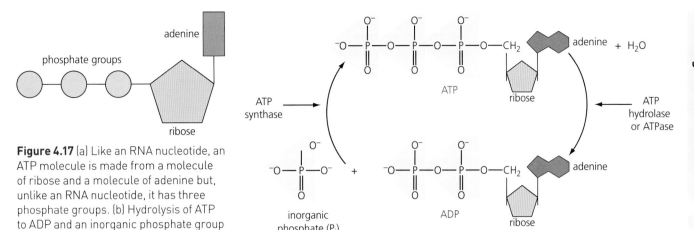

**Figure 4.17** (a) Like an RNA nucleotide, an ATP molecule is made from a molecule of ribose and a molecule of adenine but, unlike an RNA nucleotide, it has three phosphate groups. (b) Hydrolysis of ATP to ADP and an inorganic phosphate group ($P_i$) and resynthesis by condensation of ADP and $P_i$.

Processes such as protein synthesis require energy from the hydrolysis of ATP. Just like any hydrolysis reaction (see Chapter 1, page 2), water is used in a reaction catalysed by the enzyme **ATP hydrolase (ATPase)** to break the bond between the second and third phosphate groups, releasing energy. The phosphate group is often transferred to other molecules to make them more reactive so that they have the **activation energy** (see Chapter 2, page 20) to take part in another reaction. This transfer is called **phosphorylation**.

Just as quickly as ATP is used up in cells, it is resynthesised by condensing together molecules of ADP and inorganic phosphate groups ($P_i$) using another enzyme called **ATP synthase**. This requires energy. In all cells, carbohydrate or lipid molecules are hydrolysed during respiration to release the energy required to resynthesise ATP. In some cells, ATP can also be made using light energy during photosynthesis. However it is made, it is important that the cell is able to supply sufficient ATP to meet the demands of all of its energy-requiring processes, including protein synthesis.

# How genes are used

The sequence of bases in an organism's DNA carries the genetic information that determines the amino acid sequence of each polypeptide (Chapter 1, page 12) that a cell can produce. One or more very long polypeptides assemble to form proteins. Proteins include haemoglobin found in red blood cells, and enzymes that regulate all the chemical reactions of substances in cells. Because the DNA determines the sequence of amino acids in polypeptides, and hence the nature of proteins formed from them, it thereby indirectly controls how an organism develops and behaves. The fact that the genetic code that determines how information is carried on DNA is common to all living organisms and to viruses is strong evidence for evolution from a common ancestor.

The base sequence on mRNA is used by ribosomes to make polypeptide chains. Before looking at the process in more detail, it is helpful to gain an overview of how genes are used. The sequence of events leading to the production of a polypeptide from the genetic information contained in one gene occurs in two stages: **transcription** and **translation**. Table 4.5 summarises the steps in these two stages. Figure 4.18 (overleaf) is a pictorial summary of the same events in a eukaryotic cell.

**Table 4.5** The main steps in transcription and translation in a eukaryotic cell.

| Transcription | Translation |
|---|---|
| 1  The sequence of DNA bases in a single gene is used to make a molecule of messenger RNA (mRNA). | 1  mRNA is used by ribosomes to make a polypeptide chain with a sequence of amino acids encoded by the mRNA base sequence. |
| 2  At the region to be copied, the DNA unwinds. | 2  A molecule of mRNA leaves the DNA and moves to a ribosome in the cytoplasm. One or more ribosomes might attach to each mRNA molecule. |
| 3  The hydrogen bonds holding the two DNA polynucleotide chains break down, exposing unpaired DNA bases. | 3  Each ribosome travels along the molecule of mRNA, 'reading' its base sequence. As it does so, the ribosome assembles a sequence of amino acids, according to the sequence of bases on the mRNA. |
| 4  A molecule of mRNA is made which has a sequence of bases that is complementary to the base sequence of one of the DNA strands. | 4  Peptide bonds form between the amino acids brought by tRNA, making a polypeptide chain. |
|  | 5  Once finished, the polypeptide chain detaches from the ribosome. The ribosome now detaches from the mRNA molecule. |

**Figure 4.18** Transcription and translation in a eukaryotic cell.

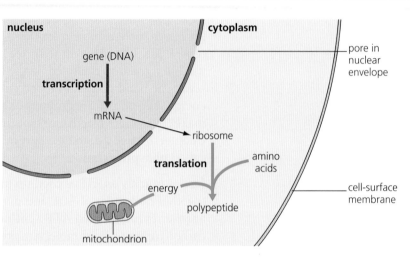

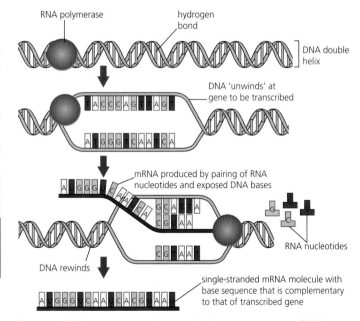

**Figure 4.19** During transcription, the base sequence of DNA is copied into the complementary base sequence of messenger RNA (mRNA). A gene is thus copied for use by ribosomes.

# Transcription: the production of mRNA from DNA

Figure 4.19 shows in more detail what happens during transcription. At the point at which a gene is to be used, the DNA molecule unwinds and the hydrogen bonds holding the two polynucleotide strands together in a DNA molecule break down. This exposes unpaired bases on the nucleotides of the two DNA strands. RNA nucleotides are already present in the cell. Bases on these RNA nucleotides form new hydrogen bonds between the exposed DNA nucleotides on *one* of the strands of DNA by a process of **complementary base pairing**. This is similar to the complementary base pairing in DNA replication, except that RNA nucleotides have the base uracil in place of the base thymine. You will read more about this on page 75.

Thus, the respective base pairs between DNA and mRNA transcribed from it are:

- adenine–uracil (A–U)
- guanine–cytosine (G–C)
- cytosine–guanine (C–G)
- thymine–adenine (T–A).

In this way, a chain of RNA nucleotides is made which has a complementary base sequence to the DNA making up one gene. An enzyme called **RNA polymerase** joins the ribose-phosphate backbone of these RNA nucleotides to form a molecule of mRNA.

Both tRNA and rRNA are made in the same way as mRNA, by using the sequence of bases in a gene. The difference is that these genes make RNAs that are used in the process of protein synthesis, whereas mRNA is translated to produce a protein. tRNA and rRNA are known as functional RNAs because they have a role in translation. tRNA carries amino acids, whereas rRNA forms part of the structure of ribosomes.

## Post-transcriptional processing of mRNA in eukaryotic cells

Earlier in this chapter you saw that not all of the DNA in eukaryotic cells codes for polypeptides or functional RNA and that the non-coding sections of DNA might be:

- between genes: these sections include DNA sequences that are repeated over and over again; they are often referred to as non-coding multiple repeats
- within genes: these non-coding sections of DNA are called **introns** and they separate the coding sequences called **exons**.

During transcription, eukaryotic cells cannot transcribe only the coding sections. Instead the whole gene, including introns and exons, is transcribed into a molecule called **pre-mRNA**. Before it leaves the nucleus, this pre-mRNA is edited. Figure 4.20 shows how this is done by removing the non-coding sections of the pre-mRNA. The coding sections are then 'edited' together to produce mRNA that carries only the coding regions of the gene, in a process called splicing.

**Functional RNA** RNA that is not translated into proteins.

**TIP**

It might help you to remember the difference between exons and introns if you think of **ex**ons as **ex**pressed sections of DNA and **int**rons as **int**ervening sections of DNA.

**Figure 4.20** During transcription in a eukaryotic cell, the entire base sequence of a gene is copied into a base sequence of mRNA. This produces a molecule of pre-mRNA that contains a copy of the non-coding regions of the DNA (introns) as well as the coding regions (exons). The non-coding regions are edited out before the mRNA leaves the nucleus.

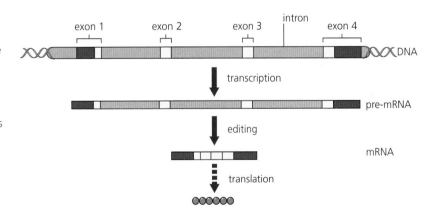

## Extension

Exons may be spliced in many alternative ways to form different mature mRNA molecules. For example, early in their development human B cells splice into one of their mRNA molecules an exon that enables a particular protein to be retained in the cell's surface membrane. Later, they stop splicing this exon and, instead, splice into their mRNA an exon that enables the protein to be released from the cell. Different splicing of the same pre-RNA at different times means that a single eukaryotic gene can actually code for more than one polypeptide chain.

## Types of nucleic acid

**1** Copy and complete the table to show the differences between the three types of nucleic acid.

| Type of nucleic acid | Hydrogen bonds present or not present? | Number of polynucleotide strands in molecule | Anticodon present or not present? |
|---|---|---|---|
| DNA | | | |
| mRNA | | | |
| tRNA | | | |

DNA is a double-stranded molecule so there are hydrogen bonds between complementary bases on the two strands. mRNA is single stranded so there is no complementary base pairing and therefore no hydrogen bonds. tRNA folds back on itself so parts of the molecule are double stranded with complementary base pairing and hydrogen bonds. Anticodons are only found on tRNA molecules.

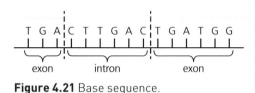

**Figure 4.21** Base sequence.

Figure 4.21 shows the bases on one strand of part of a DNA molecule.

**2** Give the sequence of bases on the mRNA transcribed from this DNA.

The intron would be removed from the pre-mRNA and the sections complementary to the two exons spliced together to form the mRNA.

## Translation

A new molecule of mRNA moves from the DNA to ribosomes in the cytoplasm. In a eukaryotic cell, this involves leaving the nucleus through one of the **nuclear pores** in the envelope surrounding the nucleus. One or more ribosomes attaches to a molecule of mRNA. Each ribosome moves along the molecule of mRNA, 'reads' the base sequence of the mRNA and uses it to assemble a sequence of amino acids. Peptide bonds form between the amino acids, creating a polypeptide. Once this polypeptide is complete, the ribosome releases the polypeptide and detaches from the molecule of mRNA.

A ribosome 'reads' the mRNA base sequence three bases at a time. Each set of three bases codes for a specific amino acid (see Table 4.6 on page 70). For this reason, each set of three mRNA bases is called a codon. Each codon is complementary to three bases, called a triplet, on the DNA from which it was transcribed.

**Codon** A sequence of three mRNA bases that codes for a specific amino acid.

**Triplet** A sequence of three DNA bases that codes for a specific amino acid.

**Anticodon** A sequence of three tRNA bases that is complementary to a codon.

Although amino acids can be free in the cytoplasm of a cell, they cannot be used by ribosomes unless they are attached to a molecule of tRNA. At the end of one of its arms, each tRNA molecule has three free bases that can base pair with a complementary codon on mRNA (Figure 4.22). At the end of another of its arms, each tRNA molecule has a site that attaches to an amino acid. The sequence of bases on the tRNA molecule that base pairs with a codon is called an anticodon. A molecule of tRNA with a particular anticodon always attaches to the same specific amino acid.

Figure 4.22 shows a single molecule of mRNA at an early stage of translation. It is attached to a ribosome and three of its bases, a codon, are shown within the ribosome. Ribosomes contain ribosomal RNA (rRNA). A molecule of tRNA is also shown. Notice that the tRNA has an anticodon that is complementary to the codon. The amino acid that is carried by the tRNA molecule is the one encoded by the codon on the mRNA, which in turn was determined by a triplet on the DNA.

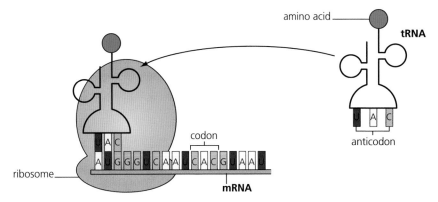

**Figure 4.22** A ribosome has attached to one end of a molecule of mRNA. The ribosome is ready to 'read' the first three bases: the mRNA codon. You can also see a molecule of tRNA that has a base sequence that is complementary to the codon; this is the anticodon of the tRNA molecule. The tRNA is attached to an amino acid.

Figure 4.23 shows a molecule of mRNA at a later stage of translation. In this diagram, you can see that the ribosome has moved along the mRNA molecule one codon at a time and has joined more amino acids together, so that a polypeptide chain is being formed by the ribosome.

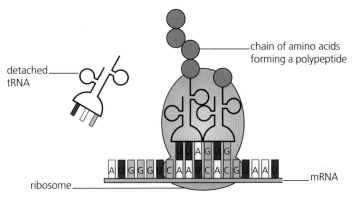

**Figure 4.23** The ribosome has moved further along the mRNA molecule and has formed a chain of amino acids that corresponds to the codons along the mRNA molecule.

Once a ribosome has 'read' the entire base sequence of a molecule of mRNA, it releases the polypeptide it has formed. The ribosome then detaches from the mRNA. It might re-attach at the start of the mRNA molecule and produce another molecule of the polypeptide. Several ribosomes can attach to, and move along, a single mRNA molecule at the same time.

## TEST YOURSELF

**11** Explain the difference between a DNA triplet and an mRNA codon.

**12** Why does transcription in prokaryotic cells produce mRNA rather than pre-mRNA?

**13** In which direction does a ribosome 'read' an mRNA molecule during translation?

**14** Explain how different polypeptides might be made from the same eukaryotic gene.

**15** Explain the difference between the genome of a cell and the proteome of a cell.

## The genetic code is common, degenerate and non-overlapping

As we have seen, organisms have coded genetic information carried on DNA. The information is carried using the genetic code, in the form of triplets. A triplet is a sequence of bases in a DNA molecule. During the production of a polypeptide, each triplet of a single gene is transcribed from DNA into a molecule of mRNA. The sequence of bases (codon) in mRNA is translated into a sequence of amino acids in a polypeptide. For this reason, the genetic code is often given as mRNA codons rather than DNA triplets.

Table 4.6 summarises the genetic code. We can use the table to show two further important features of the genetic code.

Look at the codons UAA, UAG and UGA in Table 4.6. These are 'reading instructions' for ribosomes. When a ribosome gets to these 'stop' codons it detaches from mRNA and releases its polypeptide chain.

Now look at the codons UCA, UCG, UCC and UCU. They all code for the same amino acid, serine (Ser). This means that it does not matter about the third base: UCX always codes for the amino acid serine. We describe this by saying that the genetic code is **degenerate**: several mRNA codons may encode the same amino acid.

**Table 4.6** The genetic code. Three bases (a codon) on a molecule of mRNA encode a specific amino acid. The four bases, adenine (A), cytosine (C), guanine (G) and uracil (U) can form the 64 different codons that are shown in the table. Abbreviations for the names of the amino acids that they encode are shown alongside the codons. **You do not need to remember the contents of this table!**

| | | | | | | | |
|---|---|---|---|---|---|---|---|
| AAA | Lys | CAA | Gln | GAA | Glu | UAA | STOP |
| AAG | Lys | CAG | Gln | GAG | Glu | UAG | STOP |
| AAC | Asn | CAC | His | GAC | Asp | UAC | Tyr |
| AAU | Asn | CAU | His | GAU | Asp | UAU | Tyr |
| | | | | | | | |
| ACA | Thr | CCA | Pro | GCA | Ala | UCA | Ser |
| ACG | Thr | CCG | Pro | GCG | Ala | UCG | Ser |
| ACC | Thr | CCC | Pro | GCC | Ala | UCC | Ser |
| ACU | Thr | CCU | Pro | GCU | Ala | UCU | Ser |
| | | | | | | | |
| AGA | Arg | CGA | Arg | GGA | Gly | UGA | STOP |
| AGG | Arg | CGG | Arg | GGG | Gly | UGG | Trp |
| AGC | Ser | CGC | Arg | GGC | Gly | UGC | Cys |
| AGU | Ser | CGU | Arg | GGU | Gly | UGU | Cys |
| | | | | | | | |
| AUA | Ile | CUA | Leu | GUA | Val | UUA | Leu |
| AUG | Met | CUG | Leu | GUG | Val | UUG | Leu |
| AUC | Ile | CUC | Leu | GUC | Val | UUC | Phe |
| AUU | Ile | CUU | Leu | GUU | Val | UUU | Phe |

How many other amino acids can you find in Table 4.6 that are encoded by more than one codon? Can you find any that are encoded by only one codon?

Having learned that some codons stop translation by a ribosome, you might wonder how a ribosome 'knows' where to start translating the mRNA code. This is slightly more complicated than the STOP codons in Table 4.6. In most organisms, the ribosome must 'recognise' an AUG codon that is followed by a G and has an A preceding it by three nucleotides, for example the sequence AXXAUGG.

---

**TIP**

The points you need to understand are that:

- some mRNA base sequences are start and stop messages for translation
- ribosomes only translate the mRNA molecule by taking a whole codon at a time.

---

You might realise that the sequence of mRNA bases could be 'read' in a variety of ways, but it is important that the ribosome 'reads' the sequence in only one way, so that a polypeptide with the encoded sequence of amino acids is the only one that is produced.

To explain why, let's take the following mRNA base sequence:

UCCCAUGACUCAUUCCCAGGG

If translation starts at the first codon (UCC), the order of encoded amino acids will be

serine – histidine – aspartic acid – serine – phenylalanine – proline – glycine

This results in the correct polypeptide being produced.

If the ribosome missed the first mRNA base, it would now produce a polypeptide with the amino acid sequence

proline – methionine – threonine – histidine – serine – glutamine

In other words, a completely different amino acid sequence would be produced, resulting in a polypeptide that would not have the appropriate function. Having a specific START recognition sequence ensures that the mRNA is always translated from the correct point in the mRNA molecule.

Now suppose that the ribosome were to 'read' the first mRNA base sequence above by jumping one base along each time it had 'read' a codon. The original base sequence would now be 'read' incorrectly as UCC, then CCC, CCA, CAU, AUG, and so on. This is an overlapping sequence and, again, would result in a polypeptide with an inappropriate sequence of amino acids. Translating the first three bases as a single unit, followed by the fourth, fifth and sixth in a second unit, and so on, ensures that the mRNA code is **non-overlapping**. Each base is 'read' only once in the codon of which it is a part. As a result of this non-overlapping nature of translation, an appropriate amino acid sequence is always produced from a molecule of mRNA.

## Practice questions

**1** The diagram shows part of a DNA molecule.

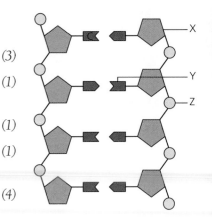

   **a) i)** Name structures X, Y and Z. (3)

   **ii)** Draw a circle round one nucleotide. (1)

   **b)** Name the bonds that form between structure X and the molecule labelled C. (1)

   **c)** Name the bonds between the nucleotides in each strand. (1)

   **d)** Describe and explain **two** features of DNA that make it useful as an information-storing molecule. (4)

**2** Below is the base sequence of mRNA that codes for part of a polypeptide in a eukaryotic cell.

ACCGUGUCCAUGUAACGU

   **a)** What is the maximum number of amino acids this sequence could code for? (1)

   **b)** Give the anticodon for the second tRNA molecule that would bind to this sequence during translation. (1)

   **c)** Write the base sequence for the DNA from which this mRNA was transcribed. (2)

   **d)** The length of the section of DNA that codes for the complete polypeptide is longer than the mRNA used to transcribe it. Give two reasons why. (2)

   **e)** The table shows the percentage of bases in two different mRNA molecules transcribed from different parts of a chromosome.

| Part of chromosome | Percentage of base | | | |
|---|---|---|---|---|
| | A | G | C | U |
| Middle | 38 | | 24 | 18 |
| End | 31 | 22 | | 21 |

   **i)** Copy and complete the table with the missing values. (2)
   **ii)** Explain why the percentages are different for the two regions of the chromosome. (2)

### Stretch and challenge

**3** This chapter began with the story of how the structure of DNA was worked out by Francis Crick and James Watson in 1953. In 1958, Francis Crick published a paper in which he described what he called the Central Dogma of Molecular Biology; that is, once (sequential) information has passed into protein it cannot get out again. This has since been simplified, perhaps oversimplified, to 'DNA makes RNA makes protein'. Since then, further discoveries about DNA and RNA have challenged the Central Dogma. Describe some of these discoveries and evaluate the extent to which you think they undermine either version of Crick's idea.

# 5 The cell cycle

## TEST YOURSELF ON PRIOR KNOWLEDGE

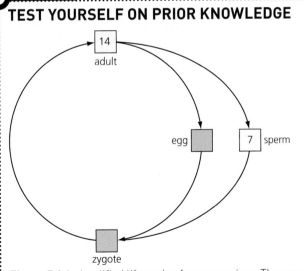

**Figure 5.1** A simplified life cycle of one organism. The numbers in the boxes represent the number of chromosomes in one cell of the organism at each stage.

1 In Figure 5.1, what are the missing numbers of chromosomes for i) egg and ii) zygote?
2 What are the components of a DNA nucleotide?
3 Name the four bases found in DNA, and identify the bases that form complementary base pairs.

## Introduction

If you have visited the Royal Botanic Gardens at Kew, in London, you probably enjoyed the variety of plants on display in the gardens and glasshouses. In addition to the staff who tend the gardens and glasshouses, there are teams of plant scientists working at the Royal Botanic Gardens.

One team works in the micropropagation unit. Members of this team are experts at cloning or propagating plants from very small pieces of plant tissue, hence 'micro'. The containers in Figure 5.2 show plants that are being grown in a special medium containing all the substances they require. They are kept in a sterile environment to make sure that the seedlings do not become infected by moulds, which could kill them. Some plants are grown in very large numbers. Often, they are plants of species from all over the world that are rare and possibly in danger of becoming extinct.

**Figure 5.2** A micropropagation unit. Inside each of the containers is a tiny plant that is growing on a sterile growth medium. Some of the plants are grown from seed. Others are grown from small pieces of growing tissue taken from a mature plant.

Cloning plants in this way relies on taking tiny clumps of dividing cells from parts of the plant that are growing. New buds are often used. Each of the cells in the clump has been formed by **mitosis** and contains all the genetic information needed to form roots, stems and flowers in the developing new plant. In this chapter, we will examine how cells pass copies of their genetic information from cell to cell and about the type of cell division that produces new, genetically identical cells.

## Replication of DNA

In Chapter 4, we saw that DNA contains two polynucleotide strands that are held together by hydrogen bonds between complementary base pairs. You will see that this base pairing is vital during **DNA replication**; that is, when DNA is copied. We can describe DNA replication using the three stages shown in Figure 5.3.

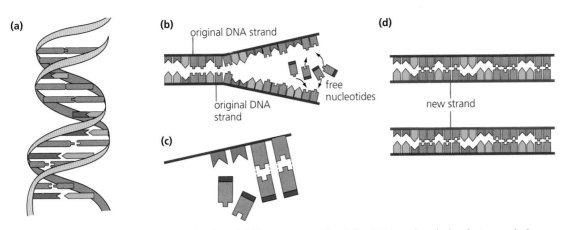

**Figure 5.3** The process of DNA replication. (a) The two strands of the DNA molecule begin to unwind when the weak hydrogen bonds between bases break. This happens at the point where the DNA is to be copied, creating a replication fork. (b) Free DNA nucleotides pair with complementary bases that are exposed on each strand of the unwound DNA. (c) Hydrogen bonds form between the new nucleotides and bases of each polynucleotide strand.

## Stage 1: the polynucleotide strands of DNA separate

Replication of a DNA molecule begins when its double helix partially unwinds (Figure 5.3a). As it does so, the hydrogen bonds between complementary base pairs break down. If you guessed that an enzyme is involved in breaking the hydrogen bonds, you are correct. The enzyme is called DNA helicase. The breakdown of hydrogen bonds allows the two polynucleotide strands to move apart. The point at which they are separating is called the **replication fork**. Figure 5.3b shows that we now have two separate strands of nucleotides with unpaired bases exposed to free nucleotides.

> **DNA helicase** The enzyme that separates the DNA double helix by breaking its hydrogen bonds.

## Stage 2: free DNA nucleotides pair with exposed bases on each polynucleotide strand

The bases on the polynucleotide strands do not remain unpaired for long. Individual DNA nucleotides have already been made in the nucleus. They are attracted to the exposed bases on each polynucleotide strand, and hydrogen bonds form between complementary bases (Figure 5.3c). For example, a free DNA nucleotide that includes the base adenine is attracted to an exposed thymine on one of the polynucleotide strands, and forms hydrogen bonds with it. This happens all along the unwound section of the DNA molecule. As a result, each of the polynucleotide strands of the DNA soon builds up a complementary sequence of nucleotides. Because each original strand is acting as a pattern for the assembly of a new strand, the original strands are known as template strands.

> **Template strand** A DNA strand that is being used as a pattern for the assembly of complementary bases into a new strand.

## Stage 3: the new nucleotides bond together

You can see in Figure 5.3c that new nucleotides have formed hydrogen bonds with the bases of each polynucleotide strand. However, these new nucleotides are not joined together themselves. This happens in the final stage of DNA replication, shown in Figure 5.3d. In this stage, phosphodiester bonds are formed between the nucleotides by condensation reactions (see Chapter 4, page 58). This linking of DNA nucleotides is controlled by the enzyme DNA polymerase.

> **DNA polymerase** The enzyme that joins free nucleotides to form a new DNA strand.

## Extension

Like all enzymes, DNA polymerase is highly specific (see Chapter 2). Because the two original strands are anti-parallel, the two new strands are also anti-parallel (see Chapter 4). The DNA polymerase can form phosphodiester bonds on one new strand as the replication fork opens up and free nucleotides pair up with exposed bases on the template strand. This is called **continuous replication**. But on the other strand, DNA polymerase enzymes must allow the replication fork to open up a short distance before being able to link new nucleotides together because the new strand is pointing the other way. This is called **discontinuous replication** because on this side the process creates the new strand in a series of short sections. These then have to be joined together by other enzymes.

The strand formed by discontinuous replication takes longer to form because the DNA polymerase keeps having to 'catch up' after the replication fork has opened further. For this reason, this strand is called the **lagging strand**, in contrast to the one being built smoothly in the opposite direction, which is called the **leading strand**.

We end up with two new DNA molecules, each formed from one original polynucleotide strand and one new, replicated, polynucleotide strand. We call this **semi-conservative replication**: one original strand remains intact (is conserved) and one new complementary strand is made using the old strand as a template for replication. The DNA molecules rewind as this process is completed. The Watson and Crick model of DNA structure (see page 56) supports the idea that a template strand is used for semi-conservative replication.

## Extension

In eukaryotes the DNA molecule in each chromosome is very long, so if replication started at one end and proceeded from one end to the other in a linear way it would be a very slow process. If the DNA molecule unwinds at a number of points along its length, forming more than one replication fork at the same time, replication is speeded up.

### TEST YOURSELF

1 One strand of DNA contains the base sequence ATCGACG. What will be the sequence of bases in the complementary DNA strand?
2 Name the type of reaction that is catalysed by DNA polymerase.
3 What is a template strand?
4 Name the enzyme that breaks the hydrogen bonds and unwinds the strands to separate them.
5 What type of bond is formed between the nucleotides in a new strand?

# ACTIVITY

## Experimental evidence for semi-conservative replication of DNA

You might wonder how we know that DNA replication is a semi-conservative process. After all, we cannot see DNA actually replicating. The evidence is indirect. Let's look at one experiment that provides evidence about DNA replication.

Look back to Figure 5.3 on page 75. The conserved polynucleotide strands and the new polynucleotide strands are coloured differently to help you to understand the replication process. We can do this in a diagram, but we cannot colour real DNA nucleotides to help identify them.

Scientists in one laboratory came up with a neat way of labelling nucleotides. It depends on the use of two isotopes of nitrogen. The more common isotope has 14 uncharged particles, called neutrons, in the nucleus of each nitrogen atom. The rarer isotope has 15 neutrons in the nucleus of each nitrogen atom. This makes the rarer isotope (15N) heavier than the more common isotope (14N). The difference in mass is tiny. However, there are so many atoms of nitrogen in a strand of DNA that a difference in mass can be detected.

Under laboratory conditions, a bacterium rapidly replicates its DNA and divides into two new cells. Bacteria use nitrogen-containing molecules in their growth medium to make DNA nucleotides. In this experiment, the scientists used two types of growth medium containing nitrogen; in one medium, all the nitrogen atoms were the $^{14}N$ isotope, and in the other they were all the $^{15}N$ isotope.

At the start of the experiment, the scientists grew bacteria on a growth medium in which all the nitrogen was the $^{14}N$ isotope. After many generations of bacteria, they removed DNA from a sample of the bacterial cells, put it into a liquid and spun it in a centrifuge. The DNA formed a band in the liquid, shown in Figure 5.4a.

The scientists then repeated this procedure exactly, but used a growth medium in which all the nitrogen was the $^{15}N$ isotope. The DNA extracted from bacteria in this experiment formed the band shown in Figure 5.4b.

Finally, the scientists inoculated a sample of bacteria from the medium containing only the $^{15}N$ isotope into

TIP
Learning how to evaluate material like this is an important skill, so make sure you can do this.

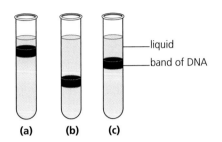

**Figure 5.4** The results of an experiment to test the theory that replication of DNA is a semi-conservative process. Each tube shows the position of DNA taken from bacteria after it was centrifuged at the same speed for the same time in a liquid. (a) DNA from bacteria grown for many generations in a medium containing only the $^{14}N$ isotope of nitrogen. (b) DNA from bacteria grown for many generations in a medium containing only the $^{15}N$ isotope of nitrogen. (c) DNA from bacteria grown for many generations in a medium containing only the $^{15}N$ isotope of nitrogen and then a single generation in a medium containing only the $^{14}N$ isotope of nitrogen.

a medium containing only the $^{14}N$ isotope. After one generation, they removed DNA from a sample of these bacterial cells and spun it in a centrifuge. This DNA formed the band shown in Figure 5.4c.

## DNA replication

Consider the experiment and answer the following questions.

1 In the final part of the experiment, the scientists inoculated bacteria from a medium containing $^{15}N$ into a medium containing $^{14}N$. Explain why.
2 The scientists removed the final sample of bacteria after only one generation in the medium containing $^{14}N$. Explain why this timing was important.
3 The scientists went on to conclude that their experiment provided evidence for the theory of semi-conservative replication of DNA. Do you agree with this? In answering this question, consider whether there is another valid conclusion from these results.

### DNA replication

Figure 5.5 shows a DNA molecule being replicated. The arrow shows the direction in which enzyme A is moving.

1 Name enzymes A and B.
*Enzyme A is DNA helicase. This is because it is the enzyme that is separating the two strands of the original DNA molecule by breaking the hydrogen bonds between the bases and unwinding the helix.*

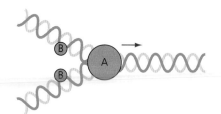

**Figure 5.5**

*Enzyme B is DNA polymerase. This is because it is the enzyme building the new strands by joining on free nucleotides as they are attracted by the exposed bases on the template strands.*

2 In eukaryotic DNA replication, a number of replication forks may open up along the DNA molecule. What is the advantage of this?
*Having replication happening at a number of points at the same time speeds up the replication of the very long DNA molecules in eukaryotic chromosomes.*

## Mitosis

According to cell theory, all cells arise from previously existing cells by cell division. Mitosis is a type of cell division that occurs in eukaryotic cells. During mitosis, a parent cell divides to produce two daughter cells. Each of the two daughter cells contains some of the cytoplasm from the parent cell. They also contain a complete copy of the parent cell's DNA, making them genetically identical to the parent cell and to each other.

On pages 74–76 in this chapter we looked at the replication of DNA. The DNA of eukaryotic cells is contained in linear chromosomes. Figure 5.6 shows you that the appearance of a chromosome changes after its DNA has been replicated. Before DNA replication, the chromosome is a single, rod-like structure containing one tightly wound double helix of DNA. DNA replication gives rise to two rod-like structures, each containing identical molecules of DNA. One region of the chromosome, the **centromere**, holds the two rod-like structures together. While they are held together they are called **chromatids**. We describe the chromatids in a pair as sister chromatids. During mitosis, sister chromatids are separated from each other. At anaphase (see Table 5.1) the chromatids become chromosomes when they separate.

**Centromere** The structure in a chromosome that holds together chromatids until they are separated by the spindle fibres.

**Figure 5.6** (a) Following the replication of DNA, a chromosome appears as a double structure composed of two chromatids. The chromatids are the products of replication of the DNA in the original chromosome. They are temporarily held together by a region called the centromere. (b) Human X (centre) and Y (lower right) sex chromosomes. Each chromosome has replicated and so there are two identical structures (chromatids) joined at the centromere.

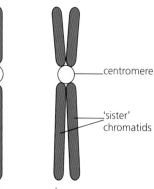

chromosome before replication

chromosome after replication

centromere

'sister' chromatids

**Table 5.1** A summary of the events occurring during each stage of mitosis.

| Stage of mitosis | Main events that occur in each stage |
|---|---|
| Interphase | Cell makes a copy of its chromosomes involving replication of DNA. Cell grows and undergoes its normal physiological functions. |
| Prophase fibre of spindle | Chromosomes coil, becoming shorter and fatter. We can now see them with an optical microscope. Nuclear envelope disappears. Protein fibres form a spindle in the cell. Each chromosome consists of two chromatids, each containing an identical DNA molecule from DNA replication. |
| Metaphase | One or more spindle fibres attaches to centromere of each chromosome. Chromosomes line up in the middle of the spindle. |
| Anaphase | Centromere holding each pair of sister chromatids together divides. Spindle fibres shorten and pull one of each pair of sister chromatids to opposite poles of the spindle. We can now refer to the chromatids as chromosomes. |
| Telophase | The two sets of separated chromosomes collect at opposite ends of the cell. A new nuclear envelope forms around each set of chromosomes. Chromosomes become long and thin. We can no longer see them clearly with an optical microscope. Cytoplasm divides to form two new cells (cytokinesis). |

Although mitosis is a continuous process, it is often described as a series of stages. The names of these stages, and the events that occur within them, are summarised in Table 5.1.

Usually, following telophase, the cell divides into two in a process called **cytokinesis**. In animal cells, the cell membrane is pulled inwards across the centre of the cell, pinching off the cytoplasm into two equal halves, each containing a new nucleus. In plant cells, vesicles fuse to extend the cell membranes across the cytoplasm and then new cell walls develop between them.

**TIP**
The terms 'chromatid' and 'chromosome' can be confusing. Chromatids are replicate chromosomes held together by a centromere. As soon as they separate, they are called chromosomes again.

**Cytokinesis** Division of the cytoplasm to give two new cells.

**TEST YOURSELF**
**6** Mitosis produces clones. What is a clone?
**7** Explain the difference between a chromosome and a chromatid.
**8** The chromosomes in Figure 5.6 were taken during prophase. How would they look different if photographed during anaphase?
**9** What happens to centromeres during anaphase of mitosis?
**10** Describe what happens during cytokinesis.

## REQUIRED PRACTICAL 2

### Preparation of stained squashes of cells from plant root tips; set-up and use of an optical microscope to identify the stages of mitosis in these stained squashes and calculation of a mitotic index

**This is just one example of how you might tackle this required practical.**

A student decided to prepare a stained squash of cells from the root tips of a plant for examining under an optical microscope. She cut off the lower part of growing roots and then placed them in ethanoic alcohol to 'fix' the tissue. This prevented mitosis from continuing in the cells. She then placed the root tips in dilute hydrochloric acid at 60°C.

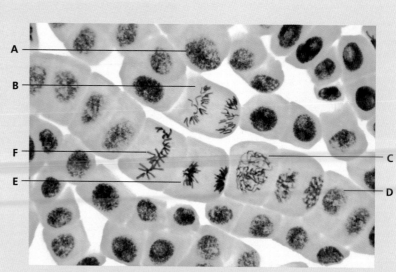

**Figure 5.7** Cells from the tip of a plant root. Many of these cells are dividing by mitosis.

This separated the cells. After this, she placed the tissue on a microscope slide with acetic orcein stain, which stained the chromosomes. Finally, she put a cover slip on top, and pressed down hard on the tissue using a folded paper towel, to obtain a layer of tissue just one cell thick. She then viewed the cells under a microscope.

Figure 5.7 shows a photograph of cells in actively dividing tissue. The chromosomes have been stained so that we can see them. Cells in this tissue were at different stages of mitosis before they were killed and stained. The cells with indistinct nuclei were not dividing at the time the photograph was taken.

1 Look at cell A in Figure 5.7. Its nucleus is clear, but all we can see is a dark-stained nucleolus surrounded by granules that are the tightly coiled regions of DNA. At which stage of mitosis was this cell?

2 Look at cell B in Figure 5.7. It has two groups of thread-like chromosomes. Look closely. Can you see that each chromosome looks V-shaped? This is because a spindle fibre, which you cannot see, is pulling its centromere to the left or right side of the cell. The arms of each chromosome lag behind its centromere, making the 'V'. At which stage of mitosis is this cell?

3 Now look at cell E. It has chromosomes that are visible but are not such clear threads as in cell B. There are two clumps of chromosomes and a clear area is developing between them. At which stage of mitosis is cell E? What is happening in the clear area between the clumps of chromosomes?

4 Cell F has a very distinctive appearance. The chromosomes are in a line across the centre of the cell. In which stage of mitosis is cell F?

5 Cells A, B, C, D, E and F are in different stages of mitosis. Put them into the correct sequence, starting with the earliest stage.

6 Is cell D in prophase or telophase? Explain your answer.

7 The mitotic index of an actively dividing tissue is found by dividing the number of cells seen in mitosis by the total number of cells counted (Chapter 14, page 245). If 320 of these plant root cells were observed and 84 were seen in various stages of mitosis, calculate the mitotic index for this dividing root tissue.

### SAFETY

Your teacher will demonstrate to you the correct procedure for focusing the microscope to avoid breaking the slide. Care should be taken when using stains – wear eye protection and make sure you follow all instructions you are given by your teacher.

### TIP

Notice how some questions were easy to answer. Anaphase is always very easy to identify. In other cases, you needed to work through a logical pathway before you could answer the question.

# The cell cycle

Not all cells in multicellular organisms retain the ability to divide. In actively dividing tissues, the new cells formed by mitosis grow before replicating their DNA and dividing by mitosis again. Thus a cycle is formed, called the cell cycle. Figure 5.8 shows this cell cycle. You can see that the two events we have described in this chapter, namely replication of DNA and mitosis, last only a short time during this cycle.

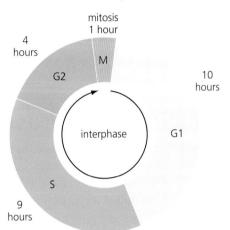

**Figure 5.8** The cell cycle in a eukaryotic cell. The times shown represent a cell cycle of 24 hours, which might be found in actively dividing tissues. The actual length of the cycle varies from one type of cell to another.

## Extension

Each part of the cycle involves specific cell activities. These are:

- G1 phase: the cell increases in size
- S phase: the cell replicates its DNA
- G2 phase: the cell increases further in size and replicates its cell organelles
- M phase: mitosis.

# Cell division in prokaryotes

When prokaryotes such as bacteria divide, the process is simpler than mitosis. Prokaryotes do not have chromosomes, just a single, circular DNA molecule. However, their plasmids are also replicated and passed on. Plasmids are really just much smaller circular DNA molecules. Bacteria can contain more than one copy of a plasmid at any one time and the maximum number of copies they can have of each is tightly regulated.

**Binary fission** The method of cell division found in prokaryotes.

Cell division in prokaryotes is called binary fission. Figure 5.9 shows how the circular DNA molecule and any plasmids in the cell undergo DNA replication. The cell then divides into two cells, each containing roughly half the cytoplasm, one copy of the circular DNA molecule and some of the plasmid copies. Prokaryotes have no membrane-bound organelles (see Chapter 3, page 51), there is no nuclear envelope to break down and there are no spindle fibres.

**Figure 5.9** (a) Binary fission taking place in bacteria. (b) The single circular DNA molecule in prokaryotes is replicated, together with any plasmids, and the two daughter cells each receive a copy of the single circular DNA molecule and a variable number of plasmids.

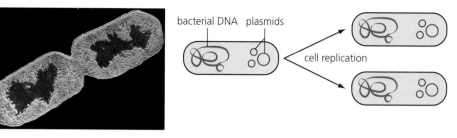

Even though there are no spindle fibres involved in binary fission, there are mechanisms to ensure that a copy of the single circular DNA molecule and some copies of each plasmid are in each half of the bacterial cell before division takes place. Obviously each daughter cell must receive a copy of the single circular DNA molecule, but even a daughter cell that fails to receive at least one copy of a plasmid may die.

# Replication in viruses

Viruses are acellular and non-living (see Chapter 6, page 99) – they do not have a cell structure, and consist only of DNA or RNA surrounded by a protein coat. This means that they do not carry out cell division. Instead, they replicate following injection of their nucleic acid into a host cell. Different viruses use either prokaryotic cells or eukaryotic animal or plant cells as hosts. Once their nucleic acid is inside a host cell, the host's DNA-replicating and protein-synthesising systems make more virus particles. Eventually, these are released when the whole host cell bursts, or by 'budding' one at a time through the host cell membrane (Figure 5.10). This is why viruses damage host cells and can cause disease.

**Figure 5.10** Virus particles leaving a host cell after replication.

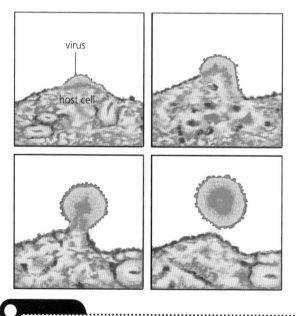

## TEST YOURSELF

**11** Name two processes involving actively dividing tissue.

**12** How many double helices of DNA are present in each chromosome (a) at the start of the G1 phase and (b) at the start of the G2 phase?

**13** Apart from the processes mentioned above, name two processes occurring in cells during interphase.

**14** List the differences between mitosis in eukaryotes and binary fission in prokaryotes.

# Cell differentiation

Not all eukaryotic cells undergo the cell cycle shown in Figure 5.8. Some fungi grow as filaments that contain cells with more than one nucleus in them. As described above, during mitosis the DNA of these fungi replicates and then the new nuclei are formed. However, the cytoplasm does not then divide. Many cells in mature plants, and most cells in mature animals, lose the ability to divide. This is why, in plant micropropagation, described at the start of this chapter, it is easier to use actively dividing cells to obtain explants.

In most multicellular organisms newly formed cells change their properties when they become specialised for specific functions. We call this process **differentiation**, and it does not occur at random. Cells in one part of a mature organism differentiate to form a tissue in which all the cells perform the same function. Tissues are organised into organs and groups of organs form systems, such as the digestive system.

# The cell cycle and cancer

During differentiation, most human cells lose the ability to divide by mitosis. Even cells that continue to divide, such as those near the surface of your skin, normally divide only about 20 to 50 times before they die. Losing the ability to divide and programmed cell death are two ways in which cells control their own division and thus numbers. Sometimes these control mechanisms break down. The result is that, if kept supplied with the necessary nutrients, such cells undergo repeated, uncontrolled division. A large mass of these cells is called a tumour. Tumours may become cancerous. Cancer occurs if cells from a tumour are able to break away and form secondary tumours elsewhere.

In about 50% of people with all types of cancer, a gene that helps to control cell growth (called *p53*) has mutated. However, there are many other reasons why the normal control of cell division breaks down. Consequently, there is no single treatment for cancer sufferers. However, many cancer treatments work by controlling the rate of mitosis.

## Extension

Cancer treatments that control the rate of mitosis do so in different ways:

- adriamycin and cytoxan are drugs that stop DNA unwinding prior to replication
- methotrexate is a drug that stops cells making DNA nucleotides
- taxol and vincristine are drugs that inhibit formation of the mitotic spindle.

## TEST YOURSELF

**15** How is a tumour formed?

**16** Which process is prevented by methotrexate? Explain your answer.

**17** Many drugs that are used to treat cancer have side effects because they also affect healthy cells. Suggest why.

# Practice questions

**1 a)** The following statements describe different stages in mitosis. Put them into the right order. *(1)*

| A | Chromatids separate and move to opposite poles of the cell. |
|---|---|
| B | Chromosomes become shorter and thicker, and the nuclear membrane breaks down. |
| C | A new nuclear membrane forms around each group of chromosomes. |
| D | Chromosomes line up along the equator of the spindle. |

**b)** Look at the figure. Put the letters A, B, C, D and E in order to indicate the correct order of the stages of mitosis. The first one has been done for you. *(1)*

C __ __ __ __

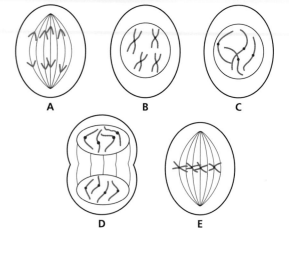

**c)** The table shows the number of cells in different stages of mitosis in part of the growing region of an onion root tip.

| Stage of mitosis | Number of cells |
|---|---|
| Interphase | 47 |
| Prophase | 18 |
| Metaphase | 4 |
| Anaphase | 2 |
| Telophase | 4 |

**i)** Calculate the mitotic index for this tissue. *(1)*

**ii)** If the whole cell cycle for these cells takes 30 hours, calculate how long the cells take to complete mitosis. *(2)*

**iii)** Calculate how long the cells spend in metaphase. *(2)*

**2** The graph shows the mean distance between the centromeres and the poles of the cell during mitosis.

**a) i)** At what time did metaphase begin? Explain your answer. *(2)*

**ii)** Calculate the mean rate of chromatid movement during the first 100 minutes of anaphase. *(1)*

**iii)** Using a tangent on the graph, find the rate of chromatid movement for the second 100 minutes of anaphase. *(2)*

**b)** Vincristine is a drug used to treat cancer. It inhibits spindle formation.

**i)** Which stage of mitosis does vincristine inhibit? Explain your answer. *(2)*

**ii)** Cancer patients who take vincristine can suffer hair loss and anaemia. Suggest why. *(2)*

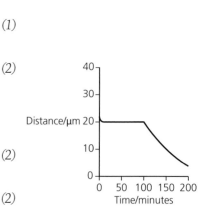

**3** Yeast cells are eukaryotic. They reproduce by a form of mitosis called budding. Scientists sampled a yeast culture every hour for 6 hours. They counted how many yeast cells were in each sample and measured the DNA concentration. The results are shown in the graph.

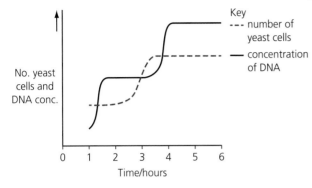

**a)** What process was happening between the 1 and 2 hour samples? *(1)*

**b)** In which stage of the cell cycle were many of the yeast cells between 1 and 2 hours? *(1)*

**c)** Between which samples was cell division happening? *(1)*

**d)** Yeast cells are single-celled organisms. Cell division by budding results in two genetically identical daughter cells in the same way that binary fission does in prokaryotes. How does budding differ from binary fission? *(1)*

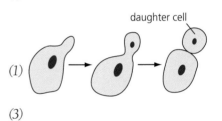

**e)** Yeast cells are eukaryotic. Using your knowledge, explain how mitosis in yeast differs from binary fission in bacteria. *(3)*

**4** The bases in DNA nucleotides contain nitrogen. Scientists grew bacteria on a medium containing $^{15}N$ ('heavy' nitrogen) for many generations. They then transferred the bacteria to a medium containing $^{14}N$ ('ordinary' nitrogen). They analysed DNA from the bacteria at three stages:

**1** while the bacteria were growing on the $^{15}N$ medium

**2** after one division of the bacteria on the $^{14}N$ medium

**3** after two divisions of the bacteria on the $^{14}N$ medium.

The DNA was analysed by extracting it from the cells and centrifuging it in a tube. The layer of DNA in the tube formed at different heights because of its density. The diagram shows their results.

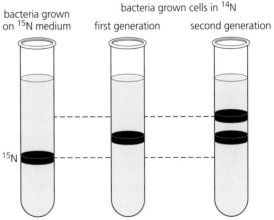

**a) i)** Copy and complete the diagram on the right to show how the tube would appear after one more division of the bacteria on the $^{14}$N medium. (2)

**ii)** Explain how these results confirm that DNA replication is semi-conservative. (3)

**b)** An alternative theory was that DNA replication is conservative. This theory suggested that the original DNA molecule remained intact and that a new molecule of DNA was produced without using any of the nucleotides from the parent strands. Complete the diagrams below to show the bands of DNA that would have appeared in the tubes if DNA replication had been conservative. (2)

After one division of the bacteria on the $^{14}$N medium.

After two divisions of the bacteria on the $^{14}$N medium.

## Stretch and challenge

**5** The replicon model was first proposed as far back as 1964 to explain prokaryotic DNA replication. What are replicons and how do they differ in prokaryotic and eukaryotic DNA replication?

# 6 The immune system

## Introduction

Look at the baby in Figure 6.1 (overleaf). He lives in the plastic cage or bubble that you can see around him. The bubble is completely airtight. The air that he breathes has to be filtered before it is pumped into the bubble. The food and water that he consumes must pass through an air lock before he can touch them. Anyone who wishes to touch him must work through the plastic gloves you can see in the photograph. This boy can never leave the plastic bubble. He will not be able to go to school. He will not be able to play out with friends. He will not be able to touch his mother, his father or his siblings. Can you imagine a life like this?

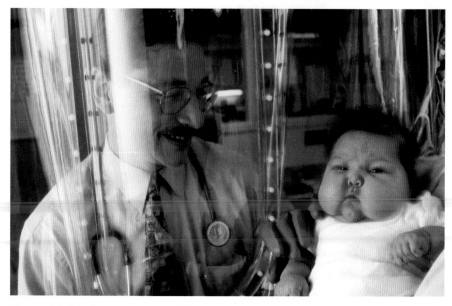

**Figure 6.1** This boy suffers from SCID. The plastic bubble in which he lives protects him from infection.

This boy was born with an inherited condition, called severe combined immunodeficiency (SCID). His immune system is unable to recognise foreign cells and does not produce enough white blood cells. White blood cells are part of the immune system, the system that helps to protect us against disease. If he were to suffer any infection he could not fight it off. As a result, even the mildest infection could kill him.

Thankfully, for most of us the body has a number of defences against pathogens. The first is to prevent their entry by various physical and chemical barriers, such as the skin, membranes, tears and saliva. If this fails, the second is for the region invaded by the pathogen to swell and become red, in what is called a non-specific inflammatory response. If this fails, the third defence occurs. Here, the body is able to recognise 'foreign' cells and target particular pathogens in a specific immune response.

Surface proteins are important in enabling the body to recognise its own cells ('self') and the cells of pathogens that are invading the body ('non-self'). Key to this is the way that proteins on the surface membranes of cells are used to identify them (see the fluid mosaic model on page 43). Understanding the role of proteins on the surface membranes of cells is vital to understanding the specific immune response. You will remember that you learned about the proteins in cell membranes in Chapter 3. Some of the surface proteins of white blood cells are receptors that bind to the proteins on the surface of pathogens. One type of white blood cell releases copies of its surface receptors as antibodies.

# Phagocytes and lysosomes

If a pathogen gets into your body, an inflammatory response is your second line of defence. This type of response is non-specific, meaning that it is the same for all pathogens. Blood contains two types of blood cell: red cells and white cells. Unlike red cells, there are many different types of white cell. Some of these are **phagocytes**. This means that they can surround and digest microscopic pathogens.

The white blood cell in Figure 6.2 is the most common type of phagocyte in your blood, called a neutrophil. You can see that its cytoplasm is full of lysosomes. You learned about lysosomes in Chapter 3 (see Table 3.1 on page 41). They contain enzymes that can digest proteins, lipids and carbohydrates and are important in destroying pathogens.

The cell-surface membrane of a neutrophil has protein receptors that bind to antigens such as those on the surface of pathogens. Their complementary binding sites enable a neutrophil to bind to a pathogen such as a bacterial cell. You can see in Figure 6.3 how a neutrophil then ingests the pathogen, forming a membrane-bound **vacuole** around it.

Lysosomes move to this vacuole and their membranes fuse with the membrane around the vacuole. This fusion releases enzymes, called lysozymes, onto the ingested pathogen. The lysozymes hydrolyse the pathogen and the neutrophil absorbs the harmless products of digestion. In this way, all phagocytes kill bacteria and other pathogens that have entered the body.

Neutrophils are carried in the blood, but they can also move and leave the blood to attack microscopic pathogens in other tissues. They leave the blood by squeezing through tiny gaps between the cells that form capillary walls (see page 158).

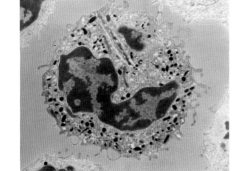

**Figure 6.2** This blood cell is a neutrophil, the most common type of phagocyte in your blood. Phagocytes are cells that can ingest and then digest microscopic pathogens. Notice that its cytoplasm is full of lysosomes (coloured red). These contain enzymes, called lysozymes, that are important in destroying ingested pathogens.

## Extension

The release of a chemical called histamine helps neutrophils leave the blood by making the walls of capillaries 'leaky'. Histamine is released by another type of white blood cell, called a mast cell, when tissues outside the circulatory system are damaged. It causes capillary walls to become more permeable so that they lose more fluid to their surroundings. This leads to localised swelling.

89

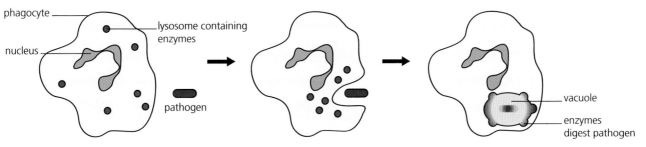

**Figure 6.3** How a phagocyte destroys a pathogen.

**TEST YOURSELF**
**1** Explain the meaning of the term 'pathogenic'.
**2** The vast numbers of bacteria living in your large intestine help to prevent infection by pathogens. Suggest how.
**3** Pathogens are harmful but, once digested by phagocytes, the products of their digestion are not. Explain why.
**4** A site of inflammation, such as a cut on your finger, becomes hotter than the surrounding skin. Suggest why.

# The specific immune response: lymphocytes

The immune system recognises and destroys any foreign cells, pathogens, abnormal cells or toxins. The immune system does this by recognising molecules on the surface of the body's own cells and identifying them as 'self'. Any cells, toxins or pathogens that have other molecules on their surface are recognised as 'foreign', and these are attacked by the cells of the immune system.

Phagocytes, which you have just learned about, do not respond in a specific way to a microscopic pathogen; they ingest and destroy any. In contrast, **lymphocytes** are specific. Lymphocytes are another type of white blood cell. Each lymphocyte attacks only one type of antigen. Figure 6.4 shows the appearance of a lymphocyte. You can see that these cells do not have the lysosomes in their cytoplasm that we have seen in phagocytes.

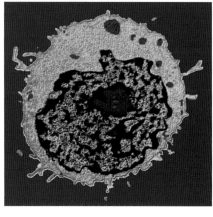

**Figure 6.4** Lymphocytes are a type of white blood cell. They are involved in cellular and humoral immunity.

Lymphocytes go through a maturing process before they are capable of fighting infection. The maturing process begins before birth, and results in two types of lymphocyte:

- B lymphocytes, known as **B cells**, mature in bone marrow; they release antibodies into the blood
- T lymphocytes, known as **T cells**, mature in a gland in the chest or base of the neck called the thymus; they cause a **cellular response** to infection and they do not release antibodies into the blood.

Before going further, we need to be clear about three terms used in immunology.

## Antigen

An antigen is a molecule that stimulates an immune response, including antibody generation. Small molecules, like amino acids, sugars and triglycerides, do not stimulate antibody generation. Antigens are large, complex molecules, such as proteins, glycoproteins and lipoproteins. Figure 6.5 shows that antigens are located on the outer surface of cells.

Each cell in your body has proteins on its surface membrane (see page 44). Normally, you would not make antibodies against these **self antigens**. In the body of another person or mammal, however, these antigens would stimulate antibody generation.

**Antigen** A large 'foreign' molecule that stimulates an immune response.

**Self antigen**

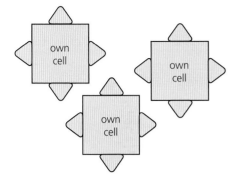

**Non-self antigens**

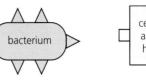

**Figure 6.5** Every mammal has self antigens on the surface membranes of its cells. The antigens on the surface of cells from another organism, non-self antigens, trigger antibody generation. (Drawings are not to scale.)

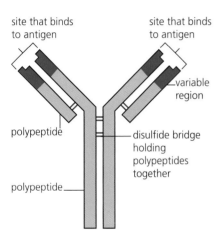

**Figure 6.6** This antibody is a Y-shaped molecule made of two long (heavy) and two short (light) polypeptides. The variable region shown in the diagram is the part that binds to an antigen to make an antigen–antibody complex.

Antibody A protein released by a B cell in response to a non-self antigen.

Antigen–antibody complex The complex formed when an antigen binds with a complementary antibody.

**TIP**
Antigens fit into the receptor site of the specific antibody. Be careful not to call this an active site, because antibodies are not enzymes.

## Antibody

An antibody is a protein released by a B cell in response to a non-self antigen. Every antibody has at least one Y-shaped molecule, made of four polypeptide chains (quaternary structure). Figure 6.6 shows one such molecule. Notice that two of the polypeptides are large and two are small. It also shows the two key parts of the antibody molecule: the sites that bind with a specific antigen. Every mammal is able to make millions of different antibodies, each with a different pair of binding sites specific to one type of antigen only.

## Antigen–antibody complex

An antigen and an antibody have complementary molecular shapes, meaning that they fit into each other. When an antibody randomly collides with a cell carrying a non-self antigen that has a complementary shape, it binds to the antigen. When this happens, the two molecules form an antigen–antibody complex. This is the first stage in the destruction of a cell carrying a non-self antigen.

Antibodies have at least two sites where they can bind to an antigen, so this means they can bind to more than one bacterium or virus. When this happens, a network of antibodies and particles forms in a clump. This is called **agglutination**. Sometimes the binding of antibodies to the antigen neutralises the pathogen. Sometimes the binding of antibodies to the antigen acts like a 'marker' which attracts phagocytic cells to engulf and destroy the cell bearing it.

# B cells and the humoral immune response

A single B cell has a unique type of receptor molecule on its surface membrane. Every day, however, you randomly make millions of B cells, each with a different receptor on its surface membrane. By chance, one of these B cells will have receptors with a shape complementary to the shape of an antigen on a cell that has entered your blood. In that case, the receptors bind to this antigen. Figure 6.7 (overleaf) summarises what then happens.

The B cell divides rapidly to produce a large number of daughter cells. Since the divisions are by mitosis (see Chapter 5), these daughter cells are

**Figure 6.7** The diagram shows what happens when, by chance, an antigen collides with a B cell. If this B cell has a complementary receptor, it will bind to the antigen. This stimulates the B cell to divide rapidly, forming a large clone of daughter cells. Most of these are plasma cells (labelled P), which release antibodies into the blood. A few are memory cells (labelled M), which remain dormant in the blood.

genetically identical; that is, they form a **clone**. The majority of cells in the clone become plasma cells, which release antibodies. A smaller number become **memory cells** and remain in circulation long after the antigen is destroyed. They can rapidly divide to produce clones of **plasma cells** if re-exposed to their complementary antigen.

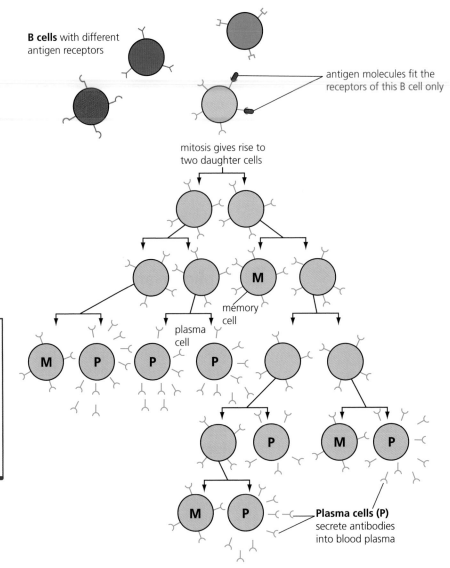

B cells with different antigen receptors

antigen molecules fit the receptors of this B cell only

mitosis gives rise to two daughter cells

M memory cell

plasma cell

Plasma cells (P) secrete antibodies into blood plasma

# T cells and the cellular immune response

Although T cells are involved in different aspects of the immune response from B cells, they respond in a similar way to exposure to a specific antigen. As Figure 6.8 shows, once it binds to its complementary antigen, a T cell divides rapidly to produce a clone of cells. T cells are, however, different from B cells in three important ways.

The receptor of a T cell has only two polypeptide chains and is never released as an antibody into the blood. T cells respond to an antigen only if this antigen is present on the surface of a macrophage that has become an **antigen-presenting cell** by presenting the antigen from an ingested pathogen, foreign cell or toxin.

**Macrophage** A type of phagocytic cell (like a neutrophil). Once a macrophage has ingested and partly digested a pathogen, it may transfer some of the pathogen's antigens to its own surface membrane. It then becomes an antigen-presenting cell.

**Figure 6.8** T cells will bind to an antigen only if it is on the surface of an antigen-presenting cell. When the surface receptor of a T cell binds to a complementary antigen, the T cell becomes sensitised and starts dividing to form a clone of cells.

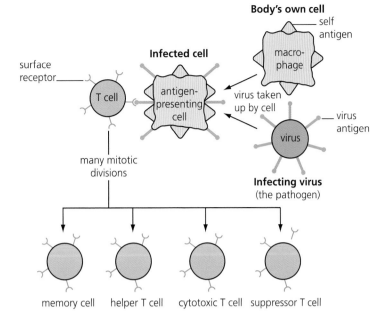

There are several different types of T cell and, following cell division, the cells in each clone differentiate into more of the same type of T cell. Cells of each type have a different function. Memory T cells remain in the blood and cause a rapid increase in the number of T cells when re-exposed to their specific antigen. Helper T cells ($T_H$ cells) assist other white blood cells in the immunological response. For example, by releasing chemical messengers, called cytokines, they stimulate:

- maturation of B cells into plasma cells that secrete antibody
- formation of memory B cells
- activation of cytotoxic T cells ($T_C$ cells), which destroy tumour cells and cells that are infected with viruses
- activation of phagocytes.

$T_H$ cells are extremely important in the immune response, and B cells cannot work without them (see pages 99–100 for what happens during infection with HIV).

---

**TEST YOURSELF**

**5** Give two ways in which antibodies are different from enzymes.

**6** A plasma cell has many mitochondria and extensive rough endoplasmic reticulum. Explain how each of these features is an adaptation for the function of this cell.

**7** After recovery from an infection, collision between a B cell and its complementary antigen is less likely than the collision between a memory cell and the same antigen. Explain why.

**8** Give two differences in the way that a $T_H$ cell and a B cell react to their respective antigens.

**9** In response to infection by a single strain of bacterium, a healthy human produces many, perhaps hundreds of, different antibodies. Explain how this happens.

# Why don't we suffer the same infection twice?

If you suffered an illness such as chicken pox when you were a child, you might have wondered why you never caught the disease again. We can explain this using our knowledge of antibodies, plasma cells and memory cells.

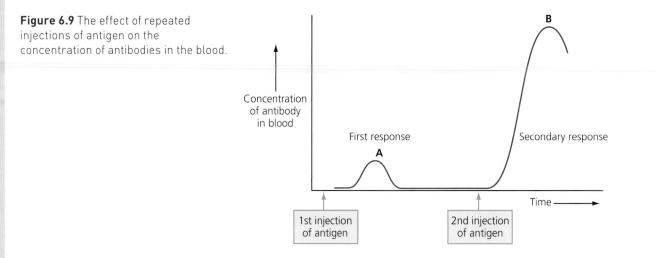

**Figure 6.9** The effect of repeated injections of antigen on the concentration of antibodies in the blood.

Look at Figure 6.9. It shows the concentration of an antibody in the blood of a person who is injected twice with the same antigen. The part of the curve labelled **A** shows what happens after the first injection. After a delay of several days, there is a small rise in the concentration of antibody in the blood. This is the **primary response**. The antibody concentration quickly falls again. We can use the information from Figure 6.9 to explain this. The short delay represents the time for a complementary, specific B cell to collide randomly with the antigen, bind with the antigen to form an antigen–antibody complex and then divide to form a clone of plasma cells that release their antibodies.

As the antibodies destroy the antigens, so fewer and fewer B cells are made, and the concentration of antibody falls again. The part of the curve labelled B shows what happens after the second injection. This **secondary response**, which may occur weeks, months or even years after the original injection, is more rapid and results in a higher concentration of the antibody in the blood. Again, we can explain this using information from Figure 6.9. After the first response, many memory cells that are specific for the antigen remain in the blood. As a result the memory cells are much more likely to collide with the antigen and bind with it. This means that the memory cells start to divide to produce plasma cells. This explains why the secondary response is more rapid than the first response. In addition, a larger number of plasma cells are produced, so the concentration of antibodies is higher during the secondary response.

The same events occur when we suffer infections. The first time we become infected by a particular pathogen, although we are able to fight the infection and recover, we produce antibodies after a time delay. During this primary response, we exhibit the symptoms of that infection. We talk about catching the disease. However, the next time we become infected by the same pathogen, our response is very rapid and we produce a large concentration of antibodies. We are able to destroy the pathogen before it causes disease.

# Why do we get colds and flu every winter?

As we have seen, the surface receptors on lymphocytes have a shape complementary to only one antigen. This is why they are a specific immune response. Memory cells can only be effective against a pathogen that always has the same antigen. If the antigen on a pathogen were to change, the receptors on our memory cells would no longer bind with it.

Some pathogens show **antigen variability**. This means that, as a result of gene mutations, they frequently change the antigens on their surface. The cold virus and the influenza virus show antigen variability in this way. This means that the 'new' cold or flu viruses have antigens that are no longer complementary to the surface receptors of the memory cells remaining from the cold or flu we caught last year. Consequently, we will not have a secondary response and so catch a cold or flu all over again.

Scientists propose explanations for the observations they make. We refer to these explanations as hypotheses. In order to test the validity of a **hypothesis**, scientists use it to make predictions and then devise experiments to test these predictions. If their results are always consistent with their predictions, they become confident in their hypothesis and it becomes a theory.

The **clonal selection hypothesis** proposes an explanation for the way that we produce antibodies against non-self antigens. The box below summarises this hypothesis.

Our immune systems produce millions of different types of B and T cells. Each type has a unique protein receptor on its surface membrane. If, and only if, one of these cells binds with a complementary antigen, it is stimulated to produce large numbers of cells that are identical to itself and, consequently, to each other. We call a group of identical cells a clone. Thus, in response to the presence of a particular antigen, a B or T cell is selected and it forms a clone. Within the clone, each cell has the identical protein receptor on its surface.

## Extension

### Testing the clonal selection hypothesis

In an experiment to test the clonal selection hypothesis, scientists injected two rats, R and S, with antigens from two different strains of pneumococcal bacteria. They injected each rat twice, with the second injection made 28 days after the first.

1 Why do you think the scientists injected each rat twice, with a gap of 28 days between injections?

*This is to allow time for the primary response to occur and memory cells to be made. After the second injection, memory cells will be present that produce very large amounts of antibody against the pathogen.*

Figure 6.10 will help you to answer this question. After the first injection, the rats would produce only a small quantity of antibodies against the pathogen (the primary response). After the second injection, the rats would produce much more of the antibody (the secondary response).

2 Which type of immune cell do you think the scientists were attempting to stimulate?

*B cells. Hopefully, you spotted that the scientists were going to look for antibodies. This means that the scientists must have been trying to stimulate B cells, since T cells do not release antibodies.*

**3** The scientists injected rat R with antigens from strain X of the bacterium (type X antigens) and injected rat S with antigens from strain Y of the pneumococcal bacterium (type Y antigens).

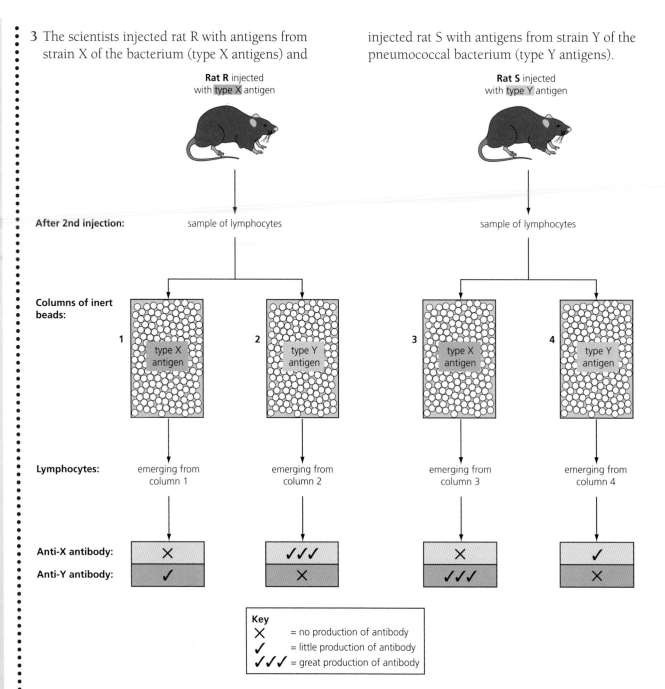

**Figure 6.10** This diagram summarises the method and results of an investigation into antibodies from rats injected with antigens.

What do you think the scientists were predicting would happen as a result of these injections?

*Using the clonal selection hypothesis, the scientists were predicting that the blood of both rats would already contain one or more B cells with a protein receptor that was complementary to each of the antigens from the two strains of bacterium. If the clonal selection hypothesis is correct, the cells in rat R with a protein receptor complementary to type X antigens should be stimulated to form a clone. However, this should not happen in rat S. Instead, the cells with a protein receptor complementary to type Y antigens should be stimulated to form a clone in rat S.*

The scientists now needed a way to find out which cells had been stimulated to produce a clone. The method they chose was rather neat. They coated inert beads with the type X antigens and put the beads into two glass columns (labelled 1 and 3 in Figure 6.10). They then did the same with type Y antigens and put these beads into another two glass columns (labelled 2 and 4 in Figure 6.10). They reasoned that if they washed samples of lymphocytes through these columns, those lymphocytes that had the appropriate complementary protein receptors would bind to the antigens on the inert beads. As a result, they would stay in the glass column and not emerge at the bottom.

4  Why was it important that the beads they used were inert?

*You might not have come across the term 'inert' before. Something that is inert will not react with anything. This was important because the scientists did not want any of the lymphocytes to react with the beads and stick to them. They wanted them to bind only with the antigens that coated the beads.*

One week after the second injection, the scientists removed a sample of blood from each of the two rats and separated the lymphocytes from the rest of the blood. They put half of each sample into a column of inert beads coated with type X antigen and half into a column of inert beads coated with type Y antigen. They then washed the samples through the columns and collected any lymphocytes that passed through the column.

5  What type of fluid do you think they would use to wash the lymphocytes through the columns?

*They would use a solution that has the same water potential as the lymphocytes.*

*Hopefully, you used your knowledge of water potential and osmosis to answer this question. Since the scientists wanted to recover some cells at the bottom of the column, they would have to use a solution with the same water potential as the lymphocytes. If they had used water, the lymphocytes would have taken up water by osmosis and burst.*

Finally, the scientists tested the lymphocytes that had passed through each column to see whether they could make antibodies against either of the two antigens used in the experiment. Figure 6.10 summarises the method they used and their results.

Now let's look at their results and see if we can interpret them. Let's start with the results from column 1. First we need to make sure we have understood what they show by describing them.

6  How would you describe the results from column 1?

*The lymphocytes washed through the column could not produce type X antibodies at all, but could produce a small quantity of antibodies against type Y antigens.*

*Remember that a description translates information from one form to another. At this stage we are not attempting to explain the result. In this case, the answer above is an adequate description.*

7  Can you explain the two parts of that description?

*The clonal selection hypothesis proposes that, by chance, rat R would have some lymphocytes with*

*protein receptors that were complementary to type X antigen, and other lymphocytes with protein receptors that were complementary to type Y antigen. So, the rat had the potential to make antibodies against both types of antigen. Since rat R was injected with type X antigen, we predict that it would produce a large clone of cells with receptor proteins complementary to that antigen. However, these would bind with the type X antigen on the beads in column 1. That is why none of the lymphocytes emerging from column 1 could produce antibodies against type X antigen (called anti-X antibody in Figure 6.10). The cells with a receptor protein complementary to type Y antigen would not bind with the type X antigen in column 1 and so emerged at the bottom of the column. The cells can produce anti-Y antibody but, since there are so few of them, they only produced a small quantity.*

Now let's describe the results from column 2. Here, the emerging cells produced large quantities of anti-X antibody but no anti-Y antibody.

8  Explain why cells emerging from column 2 give large quantities of anti-X antibody but no anti-Y antibody.

*We predicted that rat R would produce a large clone of cells with receptor proteins complementary to type X antigen. Since column 2 contained beads coated with type Y antigen, these cells would not bind to this antigen and would emerge at the bottom of the column. Since there were lots of cells, they produced a large quantity of anti-X antibodies. The few cells with protein receptors for type Y antigen bound with the antigen in column 2; none emerged at the bottom of the column.*

You should be able to use similar arguments to explain the results from columns 3 and 4. We can then evaluate the experiment. This means we ask whether the experiment was a good test of the clonal selection hypothesis. If the results had not been consistent with this prediction, they would have cast serious doubt on the hypothesis. By using two rats injected with different antigens of the same bacterium, and by using columns with only type X antigen or only type Y antigen, the scientists had built a control into their experiment. Without further details, we must assume that the scientists made sure that the conditions under which the rats were reared were kept constant. You could criticise the scientists for using only two rats, since this was a small sample size. In fact, they used large groups of rats. This account has been simplified to help us understand what was done. Therefore, we can conclude that the experiment was a valid test of the clonal selection hypothesis.

# Vaccines

Most of us are lucky not to have suffered potentially lethal diseases in order to become immune to them. You were probably given vaccinations, and as a result, your body made antibodies and memory cells. A **vaccine** is a preparation of antigen from a pathogen. The vaccine can be injected or, in some cases, swallowed. The vaccine contains antigens from the pathogen, but obviously this must be done in a safe way so that you do not become ill.

Vaccines are made harmless in a number of ways. These include:

- killing the pathogen in a way that leaves its antigens unaffected, e.g. vaccines against cholera and whooping cough
- using bacterial toxins (an antigen) to produce less harmful toxoids, e.g. tetanus, or other parts of a pathogen
- weakening the pathogen in a way that leaves its antigens unaffected, e.g. Sabin oral vaccine against polio (these weakened pathogens are said to be **attenuated**)
- using genetically engineered eukaryotic cells, such as yeast, to produce a microbial protein (an antigen), e.g. hepatitis B.

After the first treatment (primary response), you make antibodies against the antigens; you also make memory cells. After the second treatment, you show the secondary response seen in Figure 6.9 (page 94), making large numbers of B cells and memory cells. These memory cells are then able to react rapidly if a pathogen bearing the same antigen as the vaccine enters the body, killing it before it does harm.

It is very important to remember that vaccination can only work before an infection takes place. It cannot be used to treat a person who already has a disease.

# Herd immunity

Herd immunity occurs when the vaccination of a significant portion of a population provides some protection for individuals who have not developed immunity. When a high percentage of the population is protected through vaccination against a virus or bacterium, this makes it difficult for a disease to spread because there are so few susceptible people left to infect and fewer people to do the infecting. This means that the spread of disease in a population can be effectively stopped. It is particularly important for protecting people who cannot be vaccinated. These include children who are too young to be vaccinated, people with immune system problems and those who are too ill to receive vaccines (such as some cancer patients).

The proportion of the population which must be vaccinated in order to achieve herd immunity varies for each disease. However, the basic principle is that when enough people are protected, they help to protect vulnerable members in the population by reducing the spread of the disease.

However, when vaccination rates fall the herd immunity can break down, leading to an increase in the number of new cases. There was a measles epidemic in Swansea over the winter of 2012–2013. Most of the people affected were those who had not been vaccinated as babies because of the MMR scare (for more details on MMR, see pages 104–105 for the activity To vaccinate or not? A parent's dilemma). The vaccination rate in the Swansea area had fallen to below 70% of the population. As a result of the epidemic, clinics were set up in hospitals for people to receive the MMR vaccine. During the epidemic more than 1200 people became ill, 88 visited hospital and one person died.

Herd immunity can lead to the eradication of some diseases. The last case of community-acquired smallpox occurred in 1977. Scientists hope to eliminate some other diseases by vaccination, such as polio. However, this is proving more difficult.

# Viruses

## The structure of a virus

Viruses are about 50 times smaller than bacteria, and they are acellular (they do not have a cellular structure like living organisms do). They do not show any of the features of living things and only replicate when they are inside a living cell. They have a core containing genetic material, which can be either DNA or RNA but not both (see Figure 6.11). Around this there is a protein coat called a capsid. The capsid is made up of protein units called capsomeres. Some viruses have an envelope of lipid and proteins that surrounds the capsid.

## HIV: human immunodeficiency virus

The human immunodeficiency virus (HIV) is an example of a virus that is quite complex. You can see its structure in Figure 6.12. It is spherical in shape, and has an envelope made of lipids and glycoproteins. Inside this, there is a cone-shaped capsid containing RNA. It also contains an enzyme called reverse transcriptase.

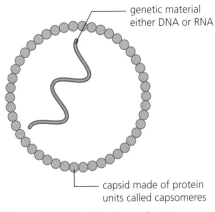

**Figure 6.11** The structure of a virus.

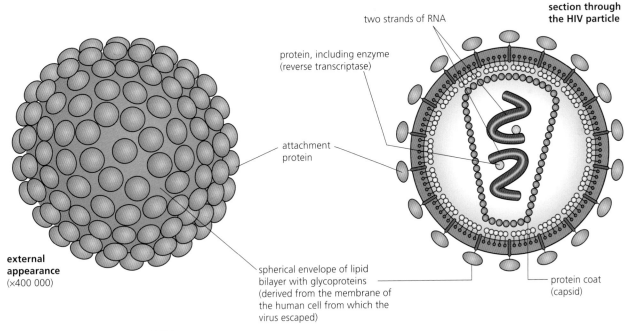

**Figure 6.12** The structure of HIV.

### How HIV is spread

HIV spreads from an infected person to another person when body fluids mix. The main ways that this can happen is during sexual intercourse, when a blood transfusion from an infected person is given to an uninfected person, when an intravenous drug abuser shares a needle with a person infected with HIV or from an infected mother to her unborn baby across the placenta.

### How HIV causes disease

HIV enters the bloodstream. It infects a type of helper T ($T_H$) cell. The viral RNA enters the cell, and the viral reverse transcriptase enzyme makes a DNA copy of the viral RNA. This DNA copy is inserted into the chromosome of the helper T cell. Every time the helper T cell divides it copies the viral DNA as well, but during this time the cell remains normal. The person is said to be HIV-positive during this stage, because they are infected with the virus and have antibodies against it in their blood.

At some point, which may be many years later, the virus DNA becomes active. It 'takes over' the cell and causes many more HIV particles to be made. This causes the helper T cell to die, releasing thousands of new HIV particles, which infect new helper T cells. Gradually the virus destroys helper T cells. You can see this in Figure 6.13.

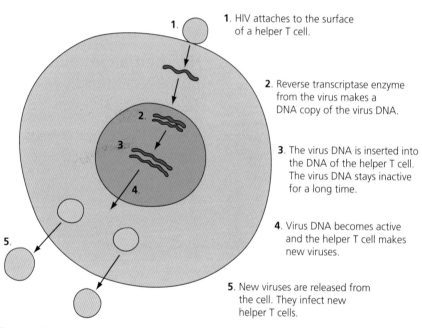

1. HIV attaches to the surface of a helper T cell.

2. Reverse transcriptase enzyme from the virus makes a DNA copy of the virus DNA.

3. The virus DNA is inserted into the DNA of the helper T cell. The virus DNA stays inactive for a long time.

4. Virus DNA becomes active and the helper T cell makes new viruses.

5. New viruses are released from the cell. They infect new helper T cells.

**Figure 6.13** How HIV infects helper T cells.

You already know the importance of helper T cells in the way the immune system works (see page 93). With a shortage of helper T cells, B cells are not activated and no antibodies are produced. This stage is called full-blown AIDS, which stands for acquired immunodeficiency syndrome. The person infected with HIV starts to suffer from diseases that would probably not cause problems in a healthy person. These are called opportunistic diseases. Examples of these are tuberculosis and Kaposi's sarcoma (a form of skin cancer that is otherwise very rare). As a result of these diseases, the person dies. There is currently no cure for AIDS, but there are now drugs that can slow down the spread of the virus within the body. With good medical care, a person with HIV can now have a life expectancy that is the same as for a non-infected person.

Antibiotics are not effective against viruses, because viruses are not living. Antibiotics are usually used against bacteria, in which they interfere with metabolism, stop cell wall synthesis, stop proteins being made at the ribosomes, and so on. Because viruses aren't cells, they have neither a cell structure nor their own metabolism, and thus antibiotics cannot affect them.

**TIP**
Be careful not to confuse **antibody** and **antibiotic**!

---

EXAMPLE

## HIV and T cells

The human immunodeficiency virus (HIV) infects and destroys only one type of T cell, the $T_H$ cells.

1 HIV can infect only $T_H$ cells. Explain why.
*This is because only $T_H$ cells have receptor proteins in their cell surface membranes that the proteins on the surface of HIV can fit into.*

People infected by this virus are HIV-positive. Without treatment, they will develop acquired immunodeficiency syndrome (AIDS). Without their $T_H$ cells, AIDS sufferers lose their ability to overcome infections. As a result, people with AIDS often die from diseases, such as tuberculosis, from which other humans could recover.

2 Without their $T_H$ cells, AIDS sufferers lose their ability to overcome infections. Explain why.
*$T_H$ cells are needed to stimulate antibody production by B cells. Without $T_H$ cells, the person's ability to produce antibodies against infections is seriously reduced.*

In 2013, the World Health Organisation (WHO) estimated that, worldwide, about 34 million people are either HIV-positive or suffering from AIDS. About 69% of these people live in sub-Saharan Africa.

Some people are at greater risk of HIV infection than others because of their lifestyles. People in 'high-risk' groups include intravenous drug users who share needles and people who have unprotected sex with many partners. Unprotected sex means sexual intercourse without using a condom, which would prevent the virus passing from partner to partner. However, scientists have found to their surprise that in many 'high-risk' groups there are a few people who seem to be resistant to HIV infection.

3 Drug users who share needles are at increased risk of becoming HIV-positive. Explain why.
*Used needles will have traces of blood in them. If the person who used the needle is infected with HIV, the virus will be present in this blood. The viruses will then be injected into the blood of the next person who uses the needle.*

Medical investigators studied a group of prostitutes at a special clinic in Nairobi, the capital of Kenya. The group included an unusually large number of HIV-resistant women. The investigators tested the blood of all the prostitutes for the presence of HIV.

4 Suggest why the investigators chose to work
a) with prostitutes
b) in sub-Saharan Africa.
*The investigators would have chosen to work with prostitutes as they are at high risk of HIV infection because they have sex with many different people. Working in sub-Saharan Africa means that this is an area where many people are infected with HIV.*

**10** Suggest reasons why smallpox could be eliminated by vaccination, but other diseases such as HIV are more difficult.
**11** Explain how a person with low numbers of helper T cells is more likely to suffer from infections.
**12** Explain how the proteins on the envelope surrounding HIV enables it to infect helper T cells.
**13** Suggest ways in which the spread of HIV can be prevented.
**14** One kind of anti-HIV drug is similar in shape to a DNA nucleotide. However, it does not contain adenine, thymine, cytosine or guanine. When the viral DNA replicates, the drug molecules are incorporated into the DNA strand. Suggest how this stops the virus replicating.

# Using antibodies

Each clone of plasma cells produces only one type of antibody, i.e. **monoclonal antibodies**. Because plasma cells can be cultured in laboratories, scientists can harvest large quantities of monoclonal antibodies. Monoclonal antibodies have a number of uses in the diagnosis and treatment of disease. For example, monoclonal antibodies against:

- a hormone, such as oestrogen, can be used to diagnose hormone deficiency
- an antigen associated with cancer, such as the prostate-specific antigen, can be used to screen for cancer (in this case prostate cancer).

Figure 6.14 shows another use of monoclonal antibodies, known as a 'magic bullet'. A drug has been attached to the molecule of antibody. Since the antibody will attach only to cells with the specific antigen on their surface, the drug will be carried directly and solely to those cells on which it is designed to work.

An example of a 'magic bullet' is a drug called T-DM1, currently undergoing trials for treatment of breast cancer. It contains the monoclonal antibody trastuzumab (Herceptin), which is already approved for treating breast cancer. This binds to receptors on the surface of some kinds of breast cancer cells. Attached to this is a toxic chemical called emtansine, which would harm too many healthy cells if it was not attached to a monoclonal antibody. There is also a third component, a chemical that keeps emtansine inactive until the monoclonal antibody has bound to a cancer cell.

## ELISA tests

Monoclonal antibodies are also useful in diagnosis. They can be used in test kits to diagnose diseases or conditions, and are very quick and reliable.

Prostate cancer is a cancer of the prostate gland, so it only occurs in men. One way to test for prostate cancer is to test the blood serum for prostate-specific antigen (PSA). If PSA concentration is abnormally high, the patient could have cancer, so further tests are carried out. The test is shown in Figure 6.15.

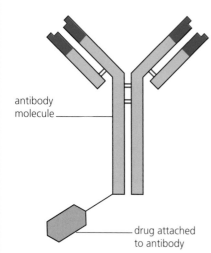

antibody molecule

drug attached to antibody

**Figure 6.14** 'Magic bullets' are monoclonal antibodies each with a molecule of a drug attached to them. These antibodies attach specifically to cells carrying the complementary antigen on their surface. In this way, the drug gets directly to the cells where it is needed.

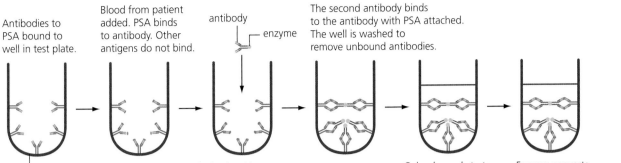

Antibodies to PSA bound to well in test plate.

Blood from patient added. PSA binds to antibody. Other antigens do not bind.

antibody — enzyme

The second antibody binds to the antibody with PSA attached. The well is washed to remove unbound antibodies.

well in test plate

Antibody with enzyme attached is added. This only binds to the first antibody if PSA is present.

Colourless substrate for enzyme added.

Enzyme converts colourless substrate to coloured product.

**Figure 6.15** Using monoclonal antibodies to test for prostate cancer.

**TEST YOURSELF**

**15** Why is the well washed after stage 4 in Figure 6.15?

**16** Why is the second antibody, with the enzyme attached, needed in the test kit?

**17** Explain the result you would get if the patient's blood did not contain PSA.

**18** Design a test kit you could use to test a patient's blood for antibodies against HIV.

Monoclonal antibodies can be used in pregnancy testing kits. As soon as the embryo implants in the lining of the uterus, a placenta starts to form. The placenta secretes a hormone called human chorionic gonadotrophin (hCG). This hormone is excreted in the woman's urine. Because the hormone is produced by the placenta, it is only found in the urine of women who are pregnant. There are several different tests available. One kind is a dipstick that is dipped into a sample of the woman's urine. If hCG is present, two blue lines will appear. The required monoclonal antibodies are all present on the dipstick. You can see this in Figure 6.16.

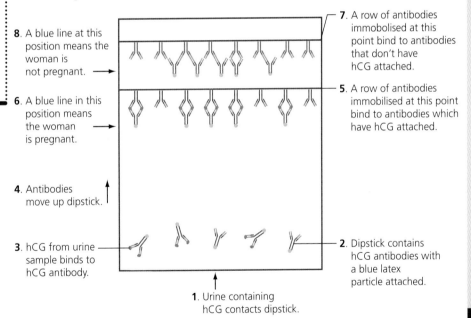

8. A blue line at this position means the woman is not pregnant.

7. A row of antibodies immobilised at this point bind to antibodies that don't have hCG attached.

6. A blue line in this position means the woman is pregnant.

5. A row of antibodies immobilised at this point bind to antibodies which have hCG attached.

4. Antibodies move up dipstick.

3. hCG from urine sample binds to hCG antibody.

2. Dipstick contains hCG antibodies with a blue latex particle attached.

1. Urine containing hCG contacts dipstick.

**Figure 6.16** How antibodies are used in a pregnancy testing kit.

**TEST YOURSELF**

**19** Explain why a pregnancy testing kit detects hCG and not any other substances present in the urine.

**20** Explain why the antibodies have blue latex particles attached to them.

**21** Explain why there are two different rows of antibodies immobilised on the dipstick.

**22** Explain how the antibodies immobilised in the two different rows are different.

# Active and passive immunity

**Active immunity** is the process that you have learned about in this chapter. It is when the immune system responds to an antigen by producing specific antibodies against the antigen, and memory B cells in a primary immune response to the antigen. This is what happens when a person is exposed to an infectious disease, or when they are given a vaccination.

**Passive immunity** is when a person is given antibodies. One example of this is when babies are breast-fed. The baby receives antibodies that their mother has made, which protects them against any infection that the mother has been exposed to. However, because the baby has not had an immune response itself, memory cells are not made so the baby has **no lasting immunity**. Another example of passive immunity is when a person is given antibodies to protect them against the effects of the venom (which is an antigen) when they have been bitten by a venomous snake.

## ACTIVITY

### To vaccinate or not? A parent's dilemma

Parents must make many decisions about the health and safety of their child. Vaccination is one of these decisions. You might think that the decision is obvious: if you have your child vaccinated, you protect it against a potentially lethal disease. However, all vaccinations carry a small risk. Suppose that you believe this risk is too high; would you have your child vaccinated?

Parents recently faced exactly this dilemma. In the 1980s, a new triple vaccine, MMR, was introduced. The MMR vaccine gives protection against three diseases:

- measles, which can lead to severe illness, convulsions, lifelong disability and death
- mumps, which can cause meningitis, permanent deafness, and sterility in males
- rubella, which during pregnancy can affect the fetus by causing deafness, blindness, heart defects and other difficulties.

The triple vaccine is thought to be a better protection than three separate vaccines because it reduces the time over which babies are exposed to rubella, measles and mumps. The MMR vaccination is made in two doses. The first dose is given at 12–15 months. The second dose is given when the child starts school.

1 The first dose of MMR coincides with the time when many breast-fed babies are weaned. What is the advantage of this timing?
2 Rubella can be passed from an infected mother to her unborn fetus. It is considered important to vaccinate boys as well as girls against rubella. Explain why.
3 Look at Table 6.1. What can you conclude from this about the safety of the MMR vaccine?

**Table 6.1** The proportion of children affected as a result of getting measles or after their first dose of a vaccine offering protection against measles.

| Condition | Proportion of children affected as a result of | |
|---|---|---|
| | ...getting measles | ...their first dose of MMR vaccine |
| Convulsions | 1 in 200 | 1 in 1000 |
| Brain disease (meningitis or encephalitis) | Between 1 in 200 and 1 in 5000 | Less than 1 in a million |
| Death | Between 1 in 2500 and 1 in 5000, depending on age | 0 |

In February 1988, Dr Wakefield, a British doctor, published a research report suggesting that MMR might cause autism, a behavioural disorder. Dr Wakefield proposed that, in some children, MMR vaccination causes inflammation of the intestine, which causes toxins to leak into the blood. These toxins then pass into the brain, producing the damage that causes autism.

4  Dr Wakefield carried out his initial research on 12 children. How reliable were his findings?

In April 2000, Dr Wakefield and Professor John O'Leary, director of pathology at a Dublin hospital, presented further research findings to the United States Congress. They reported that tests on 25 children with autism revealed that 24 of these children had traces of the measles virus in their guts. Professor O'Leary said this was now 'compelling evidence' of a link between autism and MMR.

5  Does the evidence of Dr Wakefield and Professor O'Leary show that MMR causes autism? Explain your answer.

Many other scientists performed investigations to check these research findings. None could find any evidence to support Dr Wakefield's proposal that MMR caused autism. Despite this, public confidence in the MMR vaccine fell dramatically in the UK. Many parents prevented their children from receiving the MMR vaccine. Some parents of autistic children began to sue the pharmaceutical companies that had produced the MMR vaccine.

6  Dr Wakefield acted as a consultant for some of the parents who were suing the pharmaceutical companies. This led to criticism from his scientific peers. Explain why.

7  Figure 6.17 shows the results of research in California. Do these data support the theory that autism is linked to the use of MMR vaccine? Explain your answer.

8  If everyone was vaccinated against MMR before their second birthday, there would be a correlation between the incidence of autism and being vaccinated against MMR. Explain why.

There is now an overwhelming body of evidence to suggest there is no link between MMR vaccinations and autism. Despite this, the proportion of children in the UK receiving an MMR vaccination by their second birthday fell from 91% in 1997–1998 to 81% in 2004–2005. Would you have had your baby vaccinated with MMR? We must each use the available scientific evidence intelligently to inform our decisions.

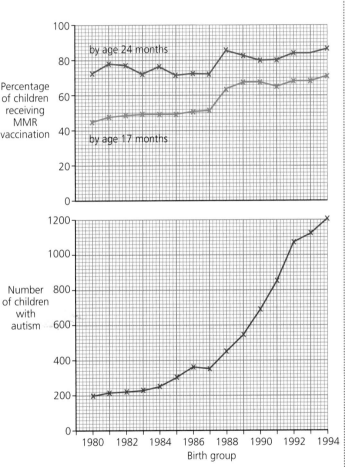

**Figure 6.17** Is there a link between the use of MMR vaccine and autism? The upper graph shows the percentage of 2-year-old children who received MMR vaccinations between 1980 and 1994. The lower graph shows the number of reported cases of autism among the children born in these years. The data are from a study in California, USA.

# Practice questions

**1 a)** The MMR vaccine contains **antigens**. What is an antigen? *(2)*

A child was given the MMR vaccine and was given a second dose of the vaccine as a booster later.

**b) i)** It took more than a week for antibodies to appear in the child's blood after the first vaccination. Explain why. *(2)*

**ii)** The concentration of antibodies increased immediately after the second vaccination. Explain why. *(2)*

**2** The diagram shows a human immunodeficiency virus (HIV).

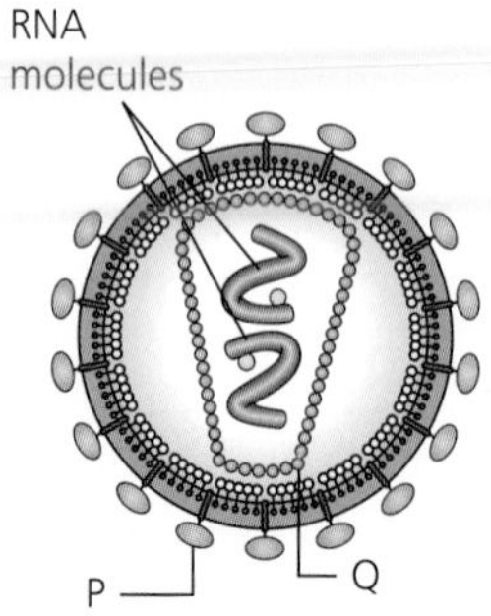

**a) i)** Name structure P and structure Q. *(2)*

**ii)** What is the function of the RNA molecules in this virus? *(1)*

**b)** Describe how new viruses are produced after HIV has infected a T cell. *(3)*

RNA molecules

P      Q

**3** A group of doctors carried out an investigation into the immune system. They injected some volunteers were injected with an antigen, A. The concentration of antibodies against antigen A in their blood was monitored for several weeks. Six weeks later, the volunteers were given a second injection containing antigen A and a different antigen, antigen B. The concentration of antibodies against both antigens in the volunteers' blood was monitored for several more weeks. The graph shows their results.

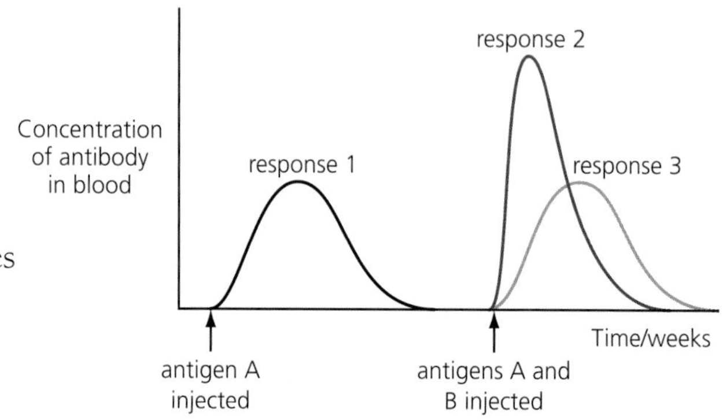

Concentration of antibody in blood

response 1

response 2

response 3

Time/weeks

antigen A injected

antigens A and B injected

**a) i)** Complete the table with ticks to indicate whether responses 1, 2 and 3 show a primary immune response or a secondary immune response. *(3)*

|  | Primary response | Secondary response |
|---|---|---|
| **Response 1** |  |  |
| **Response 2** |  |  |
| **Response 3** |  |  |

**ii)** Describe and explain the difference between response 1 and response 2. *(3)*

**b) i)** Two different responses were shown after the second injection. Explain why. *(3)*

**ii)** Suggest a reason for injecting antigen A at the same time as antigen B in the second injection. *(1)*

**4** Ibritumomab is a monoclonal antibody that has a radioactive substance attached. It is used to treat a type of cancer called non-Hodgkin lymphoma.

   **a)**   **i)** The antibody binds to cancer cells but not other cells. Suggest how. (2)

        **ii)** Explain the advantage of attaching the radioactive substance to an antibody, rather than injecting it directly. (2)

        **iii)** The monoclonal antibody is given to the patient through a drip into a vein. Suggest why it is not given as a drug to be swallowed. (2)

   **b)** A few patients who are given ibritumomab have an immune response to the drug. Explain why. (3)

**5** Read the following passage and then answer the questions that relate to it.

When it bites, a black widow spider injects venom, a toxin, into its victim. Although rarely fatal in humans, this venom causes a lot of pain that might last for several days.

There is a cure: people can be injected with antivenin, which destroys the venom from the black widow spider. The antivenin is produced by injecting the venom from black widow spiders into horses, which then produce antibodies against the venom. These antibodies are purified and stored for later use. Only people who seem near to death are normally given the antivenin, though. This is because some patients have a life-threatening reaction to the antivenin.

A new form of the antivenin, called Analatro®, has been produced and is undergoing clinical trials in the USA. Analatro is produced by injecting the venom from black widow spiders into sheep and collecting the antibodies they make. Analatro causes fewer reactions in people than the antivenin.

   **a)** Suggest the procedure the scientists would have used to ensure each horse produced enough antibodies to be used clinically. (2)

   **b)** The injected horses produce antibodies against the toxin of the black widow spider. Explain how. (4)

   **c)** Injecting people who had been bitten by black widow spiders with the horse's antibodies gave them immediate relief from the painful symptoms of the bite. Explain why. (2)

   **d)** Some people reacted badly to the injection of horse antibodies but not sheep antibodies. Suggest why. (3)

## Stretch and challenge

**6** Describe an autoimmune disease and explain how the immune system has gone wrong.

**7** Explain what happens in the case of an allergic reaction and how people are tested for allergies.

# Gas exchange

## PRIOR KNOWLEDGE

*Before you start, make sure that you are confident in your knowledge and understanding of the following points.*

- Multicellular organisms need a transport system and the complexity of this depends on the ratio of surface area to volume.
- All living organisms need a source of energy for their survival.
- In animals, the source of this energy is food.
- Food is used as a fuel in respiration to provide energy for other biochemical processes.
- The energy is released when a food compound such as glucose reacts with oxygen. The products of this reaction are carbon dioxide and water.
- Respiration is the process that takes place in cells to synthesise ATP. ATP is then used to provide energy for other biochemical processes in the cells.
- Breathing is the process by which we obtain the oxygen for respiration and release the waste carbon dioxide.
- In our lungs the oxygen enters the blood by diffusion. Diffusion is the random movement of molecules from a high concentration to a lower concentration (You learned about this in Chapter 3).

## TEST YOURSELF ON PRIOR KNOWLEDGE

1 Where does aerobic respiration take place in living organisms?
2 Name three biochemical processes in living organisms that require ATP from respiration.
3 Complete this word equation for respiration:

Glucose + ? $\rightarrow$ ? + water + energy

4 At high altitudes there is much less oxygen in the air than at sea level. When climbing high mountains (such as Mount Everest) climbers get tired and breathless and are unable to move quickly. Explain why.
5 Describe how carbon dioxide produced in the leg muscles is removed from the body.
6 If carbon dioxide is not removed from the body the blood becomes more acidic. This can be fatal. In some enclosed spaces the concentration of carbon dioxide can be quite high. For example, in grain stores the seeds release carbon dioxide, and this can be dangerous for farmers. Use your knowledge of diffusion to explain why.

# Introduction

In an emergency the first things to check are that an accident victim is breathing and that their heart is beating. These two factors are closely linked because breathing is how we obtain oxygen for respiration and the circulation of the blood distributes the oxygen around the body. Without a continuous supply of oxygen the biochemical processes that keep us alive will cease. Within a few minutes the brain will be damaged and the victim will die. Fortunately, both breathing and blood circulation are fully automated processes and we do not have to think about either. However, many of us neglect to ensure that our lungs and heart are maintained in tiptop condition and protected from damage.

# The need for gas exchange

Respiration A biochemical process by which ATP is produced, using a fuel such as glucose.

The primary function of gas exchange in animals is to supply oxygen for respiration. Respiration is the biochemical process by which ATP is produced, using glucose as a fuel. Respiration can also use other fuels, such as fats. The rate at which cells require energy varies. Active muscle cells need large supplies of energy for contraction. Cells that make digestive enzymes or that are growing rapidly use energy for the synthesis of complex molecules. The electrochemical activities of our brain cells need a constant supply of energy from respiration, and brain cells are damaged if deprived of oxygen for more than a few minutes.

Just as vital as obtaining oxygen is the removal of carbon dioxide, the waste product of respiration. Carbon dioxide produces an acid solution. As carbon dioxide accumulates the pH of the cells and blood is lowered. The pH of the blood plasma and intercellular fluid is normally maintained very close to pH 7.4. Any variation from this upsets the ionic balance, and interferes with enzyme functioning, and can rapidly lead to unconsciousness and death. To avoid this, a build-up of carbon dioxide very quickly stimulates an increase ventilation.

Gas exchange is the process by which organisms take up oxygen for respiration, and excrete the carbon dioxide produced in respiration. In this chapter we look at the structure of the gas exchange systems in different organisms and how they function. We will look at how gas exchange is affected by an organism's size and the environment in which it lives.

# Does size matter? Is big better?

A large animal is severely limited by the amount of food it can find to maintain its bulky body. This restricts the number of really large animals that can live in a particular habitat. Other problems include getting enough oxygen for respiration into the body and transporting it to where it is needed, as well as digesting and absorbing food, distributing it to all of the body cells and transporting waste to exchange surfaces, such as the kidney. Larger organisms have a greater variety of specialised cells, tissues, organs and systems, but this does not make them better than smaller organisms.

Tiny organisms are often considered simple and primitive compared with large ones. Yet it is precisely because they are simple and small that they have survived. The fact that there are now far more of them than there are large organisms shows how very successful tiny organisms have been. Each species is adapted to the particular set of conditions in which it exists, and there are many more sets of conditions suitable for small organisms than there are for large ones. Large organisms may seem to us to be dominant, more complex and more advanced, but increasing size brings a variety of problems.

# Life as a single cell

Figure 7.1 shows a single-celled organism called *Chlamydomonas*. It lives in fresh-water ponds and ditches. It is roughly spherical and about 20 μm across. It has two flagella, which enable it to swim around. It contains a chloroplast, so it can photosynthesise. Oxygen for respiration and carbon dioxide for photosynthesis are dissolved in the surrounding water and diffuse through the cell wall and the cell-surface membrane. The short diffusion pathway means that *Chlamydomonas* can rely on its cell surface for gas exchange. The maximum distance that oxygen has to diffuse to reach the centre of the cell is about 10 μm, and this takes no more than about a tenth of a second (100 milliseconds).

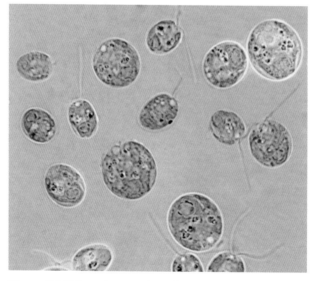

**Figure 7.1** *Chlamydomonas* is a single-celled organism that lives in water.

However, if the distance is doubled, the diffusion time is squared, so to diffuse 20 μm would take about 400 milliseconds, and to go just 1 mm would take 100 seconds. These figures are approximate because the actual rate depends on several factors, such as the concentration gradient and the material through which the oxygen is diffusing. However, it illustrates that, while diffusion is sufficiently fast to provide all parts of a small single-celled organism with enough oxygen, it is far too slow to provide enough oxygen to all parts of a larger, multicelled animal.

As the size of an organism increases, the time taken for oxygen to travel by diffusion from the cell-surface membrane to the tissues would become too great. Also, the surface area available for diffusion becomes less and less in proportion to the volume. Imagine a cube-shaped animal in which the length of each side is 1 cm (Figure 7.2). Its volume is 1 cm³, but since it has six faces each with an area of 1 cm², its surface area is 6 cm². The surface area to volume ratio is therefore 6 : 1. If you do the same calculation for 3 cm cube, the ratio is only 2 : 1. If you continue to do the calculations for larger and larger cubes, you will discover that the ratio gets smaller and smaller, so that if you were able to convert a human into a cube shape, the ratio would be less than 1 : 100. Animals are not cubic, of course, but the general rule still applies: the larger the animal the lower the surface area to volume ratio.

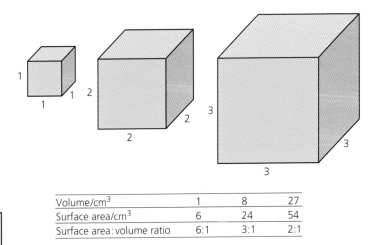

| Volume/cm$^3$ | 1 | 8 | 27 |
| --- | --- | --- | --- |
| Surface area/cm$^3$ | 6 | 24 | 54 |
| Surface area : volume ratio | 6:1 | 3:1 | 2:1 |

**Figure 7.2** This diagram shows that, as the volume of a cube increases, its surface area to volume ratio decreases.

**TIP**
See Chapter 14 to find out more about ratios.

Because it is essential to obtain oxygen and expel carbon dioxide, most animals have adaptations that increase the surface area available for gas exchange.

# How do insects breathe?

Insects may seem to be small organisms that should have little difficulty in getting enough oxygen for their needs. Many, however, are very active fliers or jumpers and have a high rate of metabolism, as anyone who has chased a buzzing fly round a room will know. A problem for insects is that they have an exoskeleton, which is fairly rigid and coated with a waxy substance. This makes it waterproof, an adaptation of a small creature living in air, to prevent it from drying out. But a waterproof surface is also difficult for gases to diffuse through. One of the reasons insects have been such an evolutionary success is that they evolved a breathing system of tubes that carry oxygen directly to all tissues and organs of their bodies.

**Spiracle** An opening in the exoskeleton of an insect that connects to the tracheal system.

Air can enter these tubes through a series of openings called spiracles arranged along the side of the body. One downside of these is that valuable water can escape. To help prevent this, the spiracles can be opened and closed using tiny valves. Some insects also have tiny hairs around the spiracles – another adaptation that reduces water loss. In addition, some insects have muscles that control ventilation of the tracheal system. This further reduces water loss.

**Tracheae** Tubes in the insect respiratory system that carry air.

As you can see from Figures 7.3 and 7.4 (overleaf), air can pass through the spiracles into a system of tracheae and narrower **tracheoles,** so that no cell is more than a short diffusion distance from a tracheole. The tracheae have rigid rings in their walls, similar to the rings of cartilage in the trachea and bronchi of humans, to keep air passages open. The tracheoles penetrate between cells and right into muscle fibres. It is here that gaseous exchange takes place. There are many of these tracheoles, giving a large collective surface area.

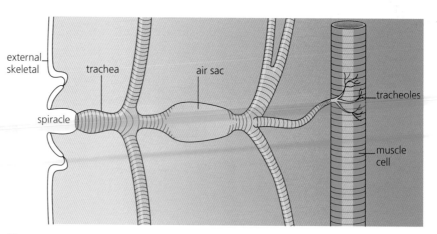

**Figure 7.3** The tracheal system of an insect.

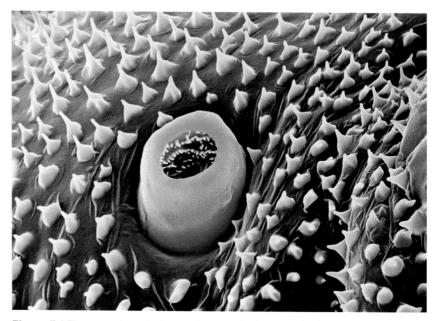

**Figure 7.4** The scanning electron micrograph shows a single spiracle in the wall of the caterpillar of the tiger moth. Along the side of the caterpillar's body is a series of spiracles through which air enters the tracheae.

In some of the tiniest insects, this system can provide enough oxygen simply by diffusion. But larger insects, such as houseflies and grasshoppers, take in oxygen more rapidly when active. This is achieved by the spiracles closing and muscles pulling the skeletal plates of the abdominal segments together. This squeezes the tracheal system and pumps the air in the sacs (see Figure 7.3) deeper into the tracheoles. The recoil also lowers the pressure inside the tracheal system so that it is lower than the atmospheric pressure outside. This results in mass flow of air into the insect. If you watch a fly or wasp closely, you can often see these pumping movements that ventilate the tracheal system.

Gas exchange between the tracheoles and tissues is illustrated in Figure 7.5.

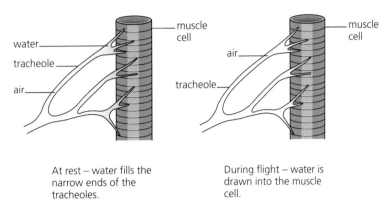

At rest – water fills the narrow ends of the tracheoles.

During flight – water is drawn into the muscle cell.

**Figure 7.5** Gas exchange between the tracheoles and tissues.

One extra adaptation helps to get additional oxygen deep into the muscles during flight. When an insect is resting, a little water leaks across the cell membranes of muscle cells. The very narrow ends of the tracheoles fill with water. When the wing muscles are working hard, they respire, partly anaerobically, and produce lactate, a soluble waste product of anaerobic respiration. This lowers the water potential of the muscle cells. As lactate builds up in the muscle cells, water passes by osmosis from the tracheoles into the muscle cells. This draws air in the tracheoles closer to the muscle cells and therefore reduces the diffusion distance for oxygen when it is most needed. This also speeds up diffusion, as diffusion is faster in gases than in liquids.

## TEST YOURSELF

**1 a)** Complete the table to show the surface area and volume of cubes of different sizes.

| Length of side of cube/cm | Surface area of cube/cm² | Volume of cube/cm³ | Surface area : volume ratio |
|---|---|---|---|
| 1 | | | |
| 2 | | | |
| 3 | | | |

**b)** How does surface area to volume ratio change as the size of the cube increases?

**2** Explain why an insect would be at risk of drying out if its exoskeleton was not covered with a waterproof substance.

**3** Suggest the function of the rigid rings in the walls of the tracheae.

**4** Suggest how the breathing system of insects helps to minimise water loss.

**5** Explain how the increase in the lactate content of the muscle cells during flight causes the removal of water from the ends of the tracheoles.

# How do fish get oxygen out of water?

**TIP**

A given volume of air contains about 30 times as much oxygen as the same volume of water.

Oxygen does not dissolve readily in water, and, as water warms up, even less can dissolve. This is bad news for fish living in lakes and rivers that are likely to get warmer as climates change. Fish are adapted to extract oxygen directly from water, unlike marine mammals such as whales and seals that have to come to the surface to take gulps of air. Not surprisingly, fish gills have a large surface area relative to the volume of the fish in contact with water through which to absorb oxygen, and this is provided by gills.

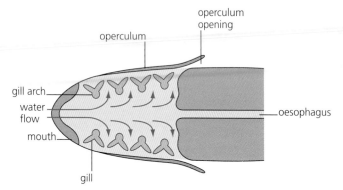

**Figure 7.6** The arrangement of the gills of a fish.

Bony fish, such as trout, perch and cod, have a series of gills on each side of the head, as shown in Figure 7.6. Each bony gill arch has two stacks of thin plates called **filaments** that stick out like leaves. On the top of each filament is a row of very thin **lamellae**, which stand out, as you can see in Figure 7.7. The surface of each lamella is a single layer of flattened cells. This covers an extensive network of capillaries so close to the surface that oxygen has only a short distance to diffuse from the water into the blood. Since fish live in water, there is no problem with such thin surfaces drying out.

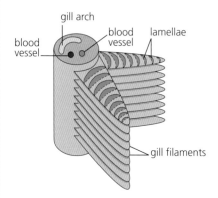

**Figure 7.7** Structure of a single gill.

The **operculum** on the side of a fish's head protects the gills from damage. Notice that the operculum opens at the rear edge. The fish takes in water through its mouth and forces it through the gills to maintain a current of water over them. The water then flows out from the back of the operculum, whether the fish is swimming slowly or rapidly.

The blood system in the lamellae is arranged so that the water flows in the opposite direction to the blood flow in the capillaries. This is called a **counter-current** system, as shown in Figure 7.8.

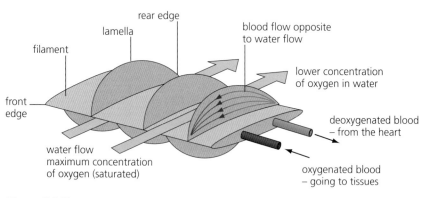

**Figure 7.8** The counter-current flow of water over a gill filament.

Generally, the concentration of oxygen in the blood is lower than in the surrounding water, so oxygen diffuses into the blood. The advantage of the counter-current system is that it maintains a diffusion gradient over the full length of the capillary. As Figure 7.8 shows, blood flows from the rear edge of the lamella to the front edge, and surrounding water flows

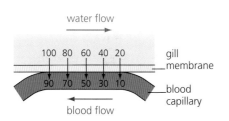

**Figure 7.9** The diagram shows that, in the counter-current flow of water over a gill filament, the gradient of oxygen concentration is maintained between water and blood. The figures show the percentage saturation with oxygen.

**Figure 7.10** Anglers use holding nets in the water to keep the fish they catch alive.

**Figure 7.11** The human breathing system.

in the opposite direction. So the blood at the front will have had the longest time for oxygen diffusion and therefore has the highest oxygen concentration. The front of the lamella is also where the surrounding water is most saturated with oxygen. At the rear edge, the blood has very little oxygen, and though there is now less oxygen in the flowing water, the diffusion gradient is about the same as at the front edge, as Figure 7.9 shows. This system therefore ensures that the concentration gradient is maintained.

### TEST YOURSELF

**6** Make a list of the features of the gas exchange system in fish, under the headings 'large surface area', 'short diffusion distance' and 'large concentration gradient'.

**7** To keep the fish alive, competition anglers keep the fish they catch in a net in water (Figure 7.10). Their catch is weighed before being safely returned to the river after the competition. Suggest why fish cannot breathe in air, even though there is a much higher percentage of oxygen in air.

## Gas exchange in humans

### The structure of the lungs

Our lungs almost fill the **chest cavity**, which occupies the upper part of the body enclosed by the ribs. The heart and its major blood vessels tuck in between the lungs and above the diaphragm that separates the chest cavity from the abdomen. The main parts of the human breathing system are shown in Figure 7.11.

115

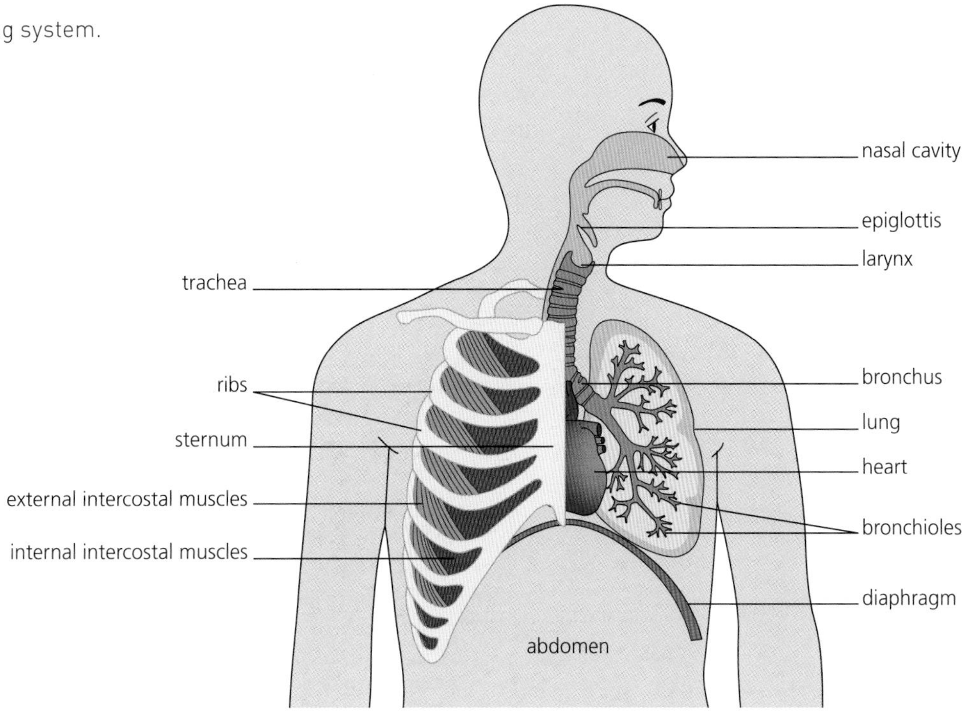

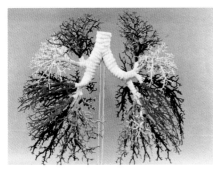

**Figure 7.12** The branching network of bronchioles in a lung. The bronchioles have been filled with plastic resin and then the other tissue of the lung has been dissolved away.

**(a)**

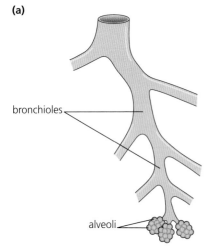

**Figure 7.13** The diagrams show that the bronchioles have narrow branches, with each branch ending in a cluster of alveoli surrounded by capillaries.

**Figure 7.14** An alveolus and blood capillaries. Note that the cells of the alveolus and the capillaries are flattened. This reduces the distance for gas diffusion.

# How oxygen gets into the blood

For efficient diffusion, an exchange surface has:

- a large surface area, compared with the volume of the organism
- a short distance for the gas to diffuse
- a large difference in the concentration of gas on opposite sides of the surface.

In human lungs, the large surface area is achieved by having a vast number of very small alveoli. Each lung contains about 350 million alveoli. As shown in Figure 7.12, the airways in the lung, called bronchioles, branch hundreds of times so that the diameter at their ends is tiny. At the end of each branch is a cluster of alveoli. These are rather like bunches of tiny, hollowed-out grapes connected to the network of bronchioles.

**(b)**

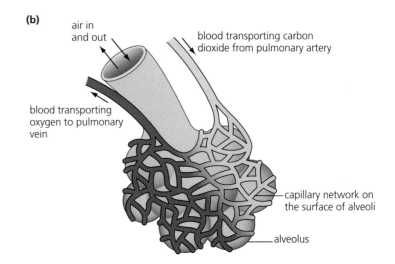

# The gas exchange surface

The alveoli are covered with a network of blood capillaries (see Figure 7.13). The walls of the alveoli and of the capillaries form a very thin barrier between the air in the alveoli and the blood (see Figure 7.14). The alveolar wall cells are flattened, with only a thin layer of cytoplasm between their cell-surface membranes. The capillary walls also consist of very thin cells. These cells are curved to form narrow tubes (see cross-section in Figure 7.14). The capillaries are so narrow that the red blood cells, which carry oxygen and carbon dioxide, touch the walls as the blood flows through the capillaries.

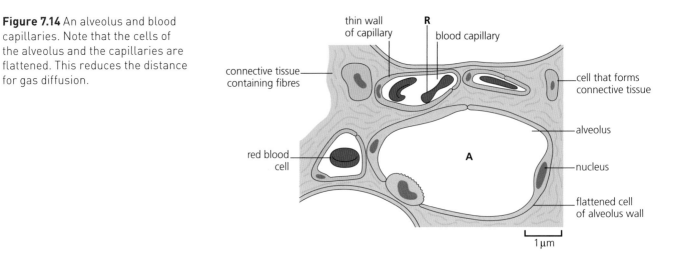

Because of this cell arrangement, the distance between the air in the alveoli and the red blood cells is very short. This minimises the distance that oxygen has to diffuse from air to blood, and carbon dioxide from blood to air.

The inner surface of the alveolus wall is covered in a very thin film of water because the cell-surface membranes of its cells are permeable to water. The rate of diffusion of a gas in water is much lower than its rate of diffusion in air, and the film of water increases the distance that gases have to diffuse. Both these factors slightly lower the rate of diffusion of gases. However, a membrane that is not permeable to water is also not permeable to oxygen. Consequently a moist surface is an unavoidable feature of a gas exchange surface. In infections such as pneumonia, the layer of liquid on the surface of the alveoli gets much thicker. This seriously slows the rate of gas diffusion.

Between the alveoli and the capillaries is a very thin layer of tissue fluid that contains elastic fibres. These fibres help to recoil the lungs, which assists in breathing out. The fibres are made by cells that fit in between the alveoli and form **connective tissue**, as seen in Figure 7.14.

A **concentration gradient** is essential for gas exchange. The gradient is maintained by ventilation in the lungs coupled with the continuous flow of blood. Breathing movements constantly result in a change of air in the alveoli, providing fresh oxygen and removing carbon dioxide. Oxygen diffuses into the red blood cells. As the blood flow rapidly moves the red blood cells on, they are replaced by oxygen-poor cells.

This ensures that the concentration of oxygen in the alveoli is always much higher than the concentration in the blood.

## TEST YOURSELF

**8** Use the scale line in Figure 7.14 to calculate the minimum distance that oxygen would have to diffuse to pass from the air in the alveolus (A) to the red blood cell (R).

**9** Smokers' lungs become lined with tar. Explain how this would affect gas exchange compared with a non-smoker.

**10** The capillaries surrounding the alveoli are so narrow that red blood cells pass through one by one, and the cells are pressed against the sides of the capillaries as they pass through. Explain how this helps efficient gas exchange.

## The structure of the breathing system

If you could take out your lungs and carefully spread out all the alveoli, they would cover most of the floor area of a typical school laboratory. That is about $70\,m^2$. The total surface area of the skin of an adult is slightly less than $2\,m^2$. The lungs therefore have an area that is roughly 35 times the area of the surface of the body. This shows how large a surface area we need for gas exchange, and it makes up for the low surface area to volume ratio of human bodies. It would obviously be impractical to have such a huge area outside our bodies, billowing out like a massive sail, or as an array of flaps sticking out from the side like external gills. Having the gas exchange surface folded away in the chest has the advantage of protecting the thin surface membrane from damage. It also reduces loss of water, as the moist air in the alveoli does not directly meet the much less moist air outside the body. As the water potential gradient is reduced, evaporation of water is also reduced.

Since the lungs are tucked away inside the chest, it is essential that a good supply of oxygen reaches the gas exchange surface. Breathing movements constantly replace the air in the alveoli by ventilation (mass transport of gases, as opposed to simple diffusion). Without active ventilation, air could not reach our lungs at a rate even close to that needed to supply our oxygen needs. Our energy demands change, for example, when we start to run. So we must also vary the rate at which we ventilate our alveoli.

Air enters our bodies through the nose or mouth and then the:

- trachea (windpipe)
- bronchi
- bronchioles
- alveoli.

### Trachea

The **trachea** is a wide tube. Air passes from the throat, through the trachea, down the neck and into the chest. Food must not go down with the air, so when we swallow a flap of cartilage called the **epiglottis** closes over the entrance to the trachea. Normally, this is precisely coordinated by a reflex action. But occasionally some food may go down the wrong way. This stimulates us to cough, which usually expels the food from the trachea and stops us from choking.

While you read, you probably lean forward and bend your neck. You do not choke since the trachea does not kink and close up as a piece of soft rubber tubing would. This is because its structure is more like a shower hose (see Figure 7.15).

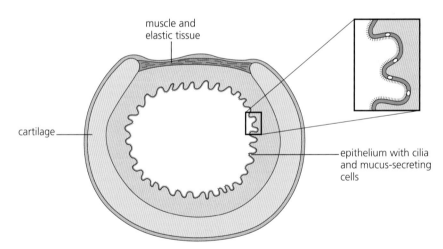

**Figure 7.15** A section across the trachea. The stiff cartilage keeps the airway open. The muscle and elastic tissue allows flexibility. Cells of the epithelium that line the airway secrete mucus, which traps bacteria and dust particles. Hair-like cilia beat upwards and sweep the mucus to the throat.

The top section of the trachea is adapted to form the **larynx** (the voice box). To produce the sounds that make up our voice, two things happen. We use precise muscular actions to adjust the position of two folds of tissue (the vocal cords) inside the larynx, and at the same time we expel air from the lungs.

### Bronchi, bronchioles and alveoli

The trachea branches into two **bronchi**, one to each lung, and these have many branches (look back at Figure 7.11). The smaller branches are called **bronchioles**, themselves repeatedly branched. The smallest branches end in the clusters of alveoli (Figure 7.13). The bronchi and the larger bronchioles also have cartilage in their walls, but here the cartilage is in small sections connected by muscle and elastic fibres. The smaller bronchioles have only muscle and elastic fibres, so that these tubes can both expand and contract easily during ventilation.

## Interpreting a photomicrograph of the lung

Figure 7.16 shows lung tissue without magnification. It appears to be quite solid and not obviously full of air. However, you can see some of the bronchioles that transport air in the lungs.

The photomicrograph in Figure 7.17 shows a very thin slice of lung tissue as seen through a microscope. At this magnification, the lung looks rather like a sponge. The sponginess is due to tiny cavities called **alveoli**. When we breathe in, the alveoli fill with air and become roughly spherical. It is in these balloon-shaped cavities that gas exchange occurs.

To obtain a slice for a microscope slide, some lung tissue is first embedded in a waxy substance that makes the tissue rigid. Then very thin slices are cut with the blade of a machine like a small bacon slicer. The alveoli appear to have irregular shapes because they are not inflated as they would be in a living lung after inhalation. The outline of their walls is wavy, not smooth and rounded. As the cells are almost transparent, the section is stained and this shows the nuclei as dark dots.

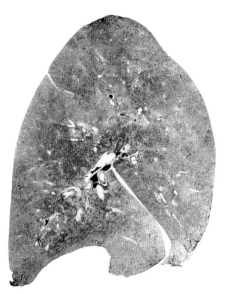

**Figure 7.16** A section of part of a lung.

**Figure 7.17** A photomicrograph of a small section of lung tissue. Since it is magnified, the spongy texture of the lung is clear.

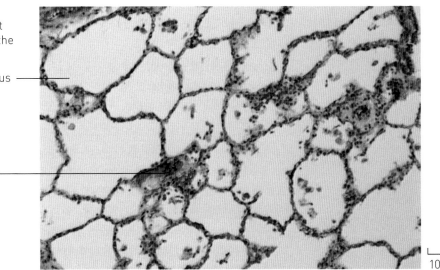

alveolus

blood vessel

100µm

119

### Analysing a photomicrograph

Remember when interpreting a micrograph of biological material that what you see through a microscope is only a very thin slice of tissue. Cutting a thin slice may alter the shape of the tissue. The slice may cut across structures in different planes (see also Chapter 3, page 39). For example, a blood vessel may be cut straight across or lengthways. Imagine if you were to cut straight across a bunch of round grapes. Some will be cut through the middle, but others will be cut to one side of the middle and these sections will look much smaller. This is the same as for cells. The tissue may be stained to show structures that would otherwise be transparent. The colours are not natural.

An optical microscope has a maximum magnification of about 1500 times. At this magnification it is impossible to distinguish structures that are less than about 2 μm apart. For example, cell-surface membranes (also called plasma membranes) are too small to be distinguished from adjacent material.

To form an image, an electron microscope uses a beam of electrons instead of light. This allows structures as small as about 1 nanometre (nm) to be distinguished. A nanometre (nm) is one-thousandth of a micrometre; 1 μm = 1000 nm. Specialised equipment is required to make a section thin enough to show structures clearly with an electron microscope. Also, in the electron microscope the specimen has to be placed in a vacuum, and this treatment can distort it.

**ACTIVITY**

### Analysing a photomicrograph

Look closely at the photomicrograph in Figure 7.17 of lung tissue.

1 Why do the alveoli appear to be different sizes?
2 How do we calculate the actual size of an alveolus from the photomicrograph?
3 What is the actual size, in millimetres, of the structure labelled alveolus in Figure 7.17?

## Breathing in

The process of breathing in increases the volume in the chest and causes air to enter the lungs. As the volume of the chest begins to increase, the air pressure inside the lungs starts to decrease. It becomes slightly lower than the atmospheric air pressure outside. This small difference in pressure causes air to move down the pressure gradient and rush into the lungs. The air movement is surprisingly rapid and forceful, yet similar small differences in pressure create the strong winds of our weather.

The volume of the chest can be increased in two ways. The chest is separated from the abdomen by a domed sheet rather like a bulging mini-trampoline. This sheet is the **diaphragm**. It is a tough membrane attached by muscles to the inner wall of the chest at the bottom of the rib cage. It seals off the chest and lungs from the organs of the abdomen. When the muscles of the diaphragm contract, the dome flattens. The centre of the diaphragm may be lowered by as much as 10 cm, thus increasing the chest volume considerably. While we are at rest only a small movement of the diaphragm is needed for us to get enough air into the lungs during each breath in.

When we are more active our oxygen requirements increase. Then, as well as the diaphragm becoming flatter, we can also move our ribs to produce a larger increase in volume. As Figure 7.18 shows, the ribs are connected to each other by two layers of **intercostal muscles**. During a deep breath in the muscles in the outer layer, called the external intercostal muscles, contract. They pull the whole rib cage upwards and outwards: each rib swings up from the backbone and its front end moves up and out (see Figure 7.18).

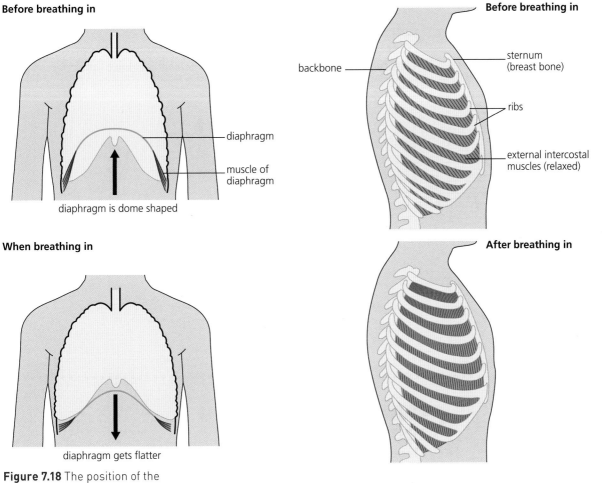

**Before breathing in**

diaphragm is dome shaped

diaphragm

muscle of diaphragm

**When breathing in**

diaphragm gets flatter

**Before breathing in**

backbone

sternum (breast bone)

ribs

external intercostal muscles (relaxed)

**After breathing in**

**Figure 7.18** The position of the diaphragm (left) and the rib cage (right), before and after taking a deep breath.

## Breathing out

At rest, breathing out is mainly due to the lungs recoiling from being stretched. As the external intercostal muscles relax, the elastic fibres round the alveoli recoil and squeeze air out. But when we are exercising it is more important to push air out forcibly. The internal layer of intercostal muscles (Figure 7.18), which slopes in the opposite direction to the external layer, contracts and helps to pull the ribs back down. We may also contract the muscles in the wall of the abdomen. This forces the liver, intestines and stomach upwards against the diaphragm, pushing it back into its domed position, so increasing internal pressure.

## Deep breathing

We do not breathe at a constant rate. When we run, our breathing gets deeper and faster. Look at Figure 7.19. It shows the changes in the volume of air in a man's lungs when he changes his activities.

**TIP**
You do not need to be able to recall the terms 'tidal volume', 'residual volume', 'vital capacity' or 'total lung capacity' in Figure 7.19.

- During the period between A and B, the man is at rest. His breathing is shallow and steady.
- At B he starts to exercise and he takes deeper breaths.
- He stops exercising at C and his breathing starts to return to its resting state.
- At D he breathes out as fully as he can by contracting his abdomen muscles so that the abdominal organs push up against the diaphragm (see page 121). He also uses his internal intercostal muscles to pull down his rib cage as far as possible. This empties his lungs much more than when breathing at rest. That still leaves quite a lot of air in the lungs that cannot be expired. This amount of air is the residual volume of his lungs.
- Then, at E, he breathes in as deeply as he can (his maximum chest expansion).

Residual volume The volume of air left in the lungs after as much air as possible is breathed out.

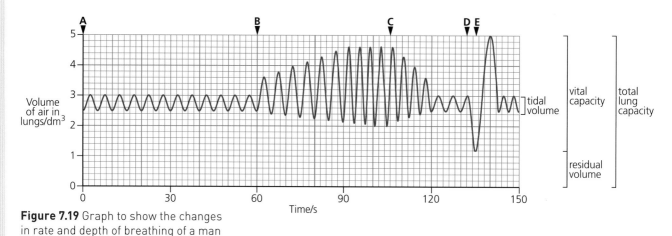

**Figure 7.19** Graph to show the changes in rate and depth of breathing of a man who changes his activities.

## Extension

The volume of air inspired per breath when at rest is the **tidal volume**. Breathing is at a steady rate and the volume taken in is the same each time. Therefore, to work out the volume of air inspired per minute, the **pulmonary ventilation rate**, simply multiply the tidal volume by the number of breaths per minute, as represented by the equation:

Pulmonary ventilation rate = tidal volume × number of breaths per minute

**TEST YOURSELF**
11 Use the graph in Figure 7.19 to answer the following questions.
   a) What is the tidal volume for the man when he is at rest?
   b) How many breaths per minute does the resting man take?
   c) Calculate the pulmonary ventilation rate of the resting man.
   d) What was the maximum volume of a single breath while exercising?
   e) Calculate the percentage increase for the maximum volume inspired during exercise compared with the resting tidal volume.
   f) The vital capacity is the maximum volume of air that can be inspired after expiring as fully as possible. What is the vital capacity for this man?
   g) What is the total lung capacity, including the residual volume, that cannot be expired?

## ACTIVITY

### Interpreting pressure changes during breathing

Look carefully at the two graphs and make sure you understand what they show.

- The upper graph shows the changes in the relative pressure in the alveoli during one breath; that is, during both inspiration and expiration. Pressure is measured in kilopascals (kPa). Zero on the *y*-axis is when the pressure in the alveoli is the same as the atmospheric pressure outside the body. Atmospheric pressure is normally about 100 kPa, but it varies according to weather conditions.
- The lower graph shows the changes in the volume of the lungs during the same breath. As lung volume increases, pressure falls and air enters the lungs. The volume is measured in cubic decimetres. Remember that 1 dm³ equals 1000 cm³.

1 **a)** What is the maximum increase in the volume of the lungs?
  **b)** Do you think these data were measured when the man was at rest or during exercise? Explain the evidence for your answer.

2 Describe the pattern of change in volume of the lungs during inspiration.

3 Describe the pattern of change in pressure in the alveoli during inspiration.

4 Explain what causes the decrease in pressure in the alveoli at the beginning of inspiration.

5 Explain why the pressure in the alveoli returned to zero at the end of inspiration.

6 Describe the pattern of change in volume in the alveoli during expiration.

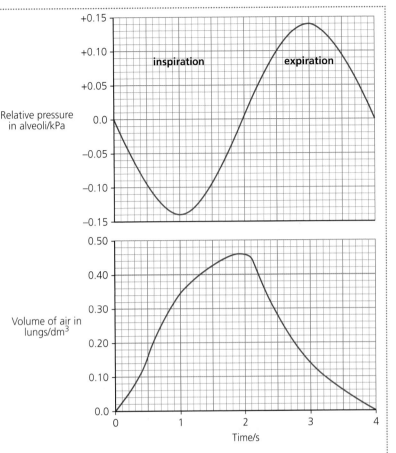

Relative pressure in alveoli/kPa

Volume of air in lungs/dm³

Time/s

**Figure 7.20** Graphs showing changes in alveolar air pressure and lung volume during inspiration and expiration in an adult man.

7 Describe the pattern of change in pressure in the alveoli during expiration.

8 Explain the changes in pressure and volume during expiration.

9 A chest wound (such as a stab or bullet wound) that allows air into the space between the chest wall and the lungs can prevent normal inspiration, even though breathing movements occur. Suggest an explanation for this.

## TEST YOURSELF

12 List in order the parts of the breathing system that carbon dioxide goes through as it passes from the blood to the air outside the body.

13 How are the lungs protected from bacterial infection?

14 Give three ways in which the alveoli are adapted for rapid diffusion of oxygen into the blood.

15 Explain how the oxygen diffusion gradient between the air in the alveoli and the blood in the surrounding capillaries is maintained.

16 Suggest why the air we breathe out has more water vapour than the air we breathe in.

17 How is the volume of the chest cavity increased during inspiration?

18 What is the 'tidal volume'?

19 How does the ventilation rate change during exercise?

## TIP

See Chapter 14 to find out about interpreting line graphs, taking measurements and describing patterns and trends.

# Smoking kills

Environmental pollutants can be difficult to avoid, but many people willingly draw tobacco smoke into their lungs, despite the evidence that this will do long-term damage. Tobacco smoke contains a vast mixture of substances. These include the addictive ingredient nicotine and the highly toxic gas, carbon monoxide. Of over 4000 organic compounds in the tar, at least 60 are known to cause cancer. All of these substances reach the delicate cell lining in a smoker's lungs. Some get swept out again when the smoker exhales, some stay in the lungs and some pass into the blood and reach all parts of the body. After a few years, a smoker's lungs are lined with tar, as the photograph in Figure 7.21 shows.

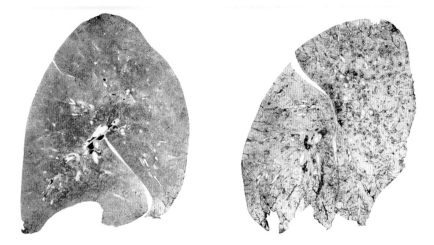

**Figure 7.21** Left: section through the clean lung of a non-smoker. Right: section through a tar-filled smoker's lung.

It is estimated that each year in the UK about 115 000 people die early due to their smoking habit. Of these, about 28 000 die from lung cancer, and over 30 000 die from other chronic lung diseases such as emphysema and bronchitis. Smoking accounts for between 80 and 90% of deaths from lung diseases. On average, of those who continue to smoke regularly from their teenage years, about a quarter die prematurely in middle age, 20 years before the normal life expectancy.

## Interpreting data relating to the effects of smoking and pollution

### Collecting the evidence

Researchers investigating the effects of smoking or other pollution cannot carry out controlled experiments on people. It is not ethically acceptable, for example, to select randomly two groups of people and make one group smoke 20 cigarettes a day for 10 years, while banning the control group from smoking altogether. Evidence must be based on people who were exposed to smoking, or another pollutant, and then comparing them with others who were not exposed.

In **retrospective studies** (collecting data from the past), researchers select groups of people who have already developed a disease such as lung cancer. They then question the people about their past experiences and look for common factors. They may also compare their experiences with the experiences of a control group who do not have the disease being studied. These studies can be unreliable because people may have forgotten details such as how long they have smoked, or they may deliberately deceive or exaggerate.

# Extension

In **prospective studies** (collecting data as it accumulates), researchers select groups and then follow what happens to them over a period of years. This makes it easier for the researchers to keep track of changes, such as changes in smoking habits, as long as frequent checks are made. The researchers can also keep records of a wider range of possible variables. However, this adds to the time it takes to get useful results.

## Interpreting the results of studies on health risks

You will understand that there are often great difficulties in collecting data on health risks.

- There are often many factors involved.
- Controlled experiments in which one variable only is studied cannot be carried out on people for ethical reasons.
- It is hard to find enough people with similar lifestyles to act as matched control groups.
- It is often several years before the effects of health risks become apparent, and following up groups of people for long periods is difficult and expensive.
- Data obtained by asking people about their past are often unreliable.

When looking at the results of a health risk study, you need to consider the following.

- Find out the number of people who were investigated. You can have more confidence in the evidence if a large number of people were involved than if the number was small.
- Identify the different levels of exposure to the health risk that were investigated; for example, the number of cigarettes smoked per day.
- Assess whether the control group is well matched with the group exposed to the factor being tested. They should, for example, come from similar backgrounds, be of similar ages and so on.
- Assess whether the differences between the results for the two groups are sufficiently large to indicate that the factor thought to be a health risk is indeed a risk.
- Find out whether tests have been done to check that the differences are statistically significant.

If the number of cases (incidence) of the disease or the number of deaths (mortality rate) is given, calculate the **relative risk**; for example, calculate the *difference in the percentages* of the two groups that develop the disease.

## Terms used in investigations of the causes of diseases

- **Incidence** The incidence of a disease is the number of cases that occur in a particular group of people in a given time, such as the number of smokers that develop lung cancer in a year. To make it easy to compare how common a disease is in different groups, the incidence is calculated as the number of cases in a standard size of group, for example, number of cases of lung cancer per 1000 smokers per year. When looking for possible effects of pollution on asthma, researchers might compare the incidence of asthma in children living in urban areas with those in rural areas.
- **Mortality rate** The mortality rate is the number of deaths per number of the population per year from a particular disease or other cause, such as

road accidents. A rate of 2.5 lung cancer deaths per 1000 smokers would mean that, out of 10 000 smokers, on average 25 died each year.

- **Correlation** A correlation is an association between two variables. If measurements of both variables increase, there is a positive correlation. For example, there is a well-established correlation between the number of cigarettes smoked per day and the incidence of lung cancer. It is important to note that a positive correlation is *not* proof that one factor is the cause of another. You might find a correlation between baldness and wearing a hat, but that would not mean that wearing a hat causes baldness.

- **Statistically significant** Investigations of the effects of smoking, pollution, drugs or diet on people never give absolutely certain results. You do not find, for example, that everyone who smokes 30 cigarettes a day for 5 years gets lung cancer. People are highly variable and do not always behave consistently. Researchers carry out statistical tests on their results to find out whether it is likely that that they have discovered a genuine effect. These tests check how likely it is that a difference, for example, between smokers and non-smokers, is just due to chance. This will depend on the number of people tested and the size of the differences in results for the two groups. The difference between the results is usually considered to be statistically significant if it shows a more than 95% probability that the difference is real (not due to chance). Even so, this is not absolute proof. There is still a 1 in 20 probability that these particular results were due to chance.

- **Risk factor** A risk factor is something, such as smoking, that correlates with an increased chance of suffering from a particular disease or condition. The relative risk can be calculated by finding the ratio between the incidence of the disease in those exposed to the factor and the incidence in those not exposed. For example, if the incidence of a lung disease in smokers is 30 cases per 1000 per year and in non-smokers it is 10 cases, the relative risk is 3.0, which means that smokers are three times more likely to develop the disease.

## What is the evidence for the link between smoking and disease?

Tobacco was introduced into Britain over 400 years ago. It was another 300 years before smoking cigarettes began to be both popular and affordable. By the 1930s, doctors were beginning to suggest that smoking might be damaging to health as they were seeing greatly increased numbers of patients suffering from lung cancer. Their concerns did not affect the smoking behaviour of the general public. The graph in Figure 7.22 shows how the consumption of cigarettes in the UK grew during the twentieth century.

**Figure 7.22** Graph of cigarette consumption by men and women in the UK from 1900 to 1980.

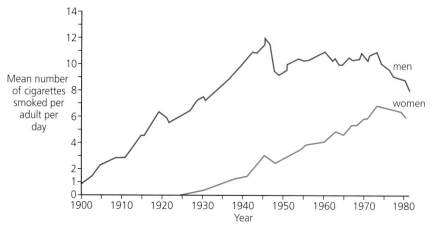

The graph in Figure 7.23 shows the changes in the number of deaths from lung cancer up to 1960 compared with cancers of other organs.

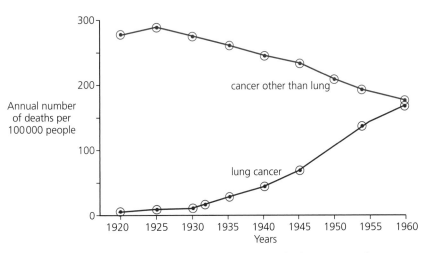

**Figure 7.23** Death rates from cigarette lung cancer and other cancers between 1920 and 1960.

Look at Figure 7.22. Notice that the number of cigarettes smoked by men rises steadily from 1900. How does this compare with the number of deaths from lung cancer, shown in Figure 7.23? This also rises, but the rise only becomes steep after about 1930. Then, both the number of cigarettes smoked and the number of lung cancer deaths go up steeply, so there is a **positive correlation** (see page 247) between them.

How could the delay in the rise in lung cancer deaths be explained? Bear in mind that cancer develops only slowly, so it may be 20 years or more before smokers start to show signs of the disease and die.

It is important to note that this positive correlation could not prove that smoking causes lung cancer. Many other factors were also increasing over the same period (such as the number of cars), and it was possible that one of these factors could explain the increase in lung cancer. The correlation did, however, provide one piece of evidence that supported the hypothesis and justified further research.

Much more evidence was needed to prove that smoking really does cause lung cancer. In 1950, two epidemiologists, Richard Doll and Austin Hill, published the results of a study of patients in four hospitals in the UK. Two groups of patients were selected: those who were suffering from lung cancer and those who had been admitted to hospital for other conditions. The patients were questioned about their smoking habits. They were, for example, asked whether they were regular smokers or had given up smoking, and about their daily consumption of cigarettes. Doll and Hill showed that a higher proportion of the lung cancer patients than the control group were regular smokers. Although other studies were showing similar results, they were still not considered to be absolute proof of a link between smoking and lung cancer. Critics suggested that some unknown factor might make people both more likely to take up smoking and to be susceptible to lung cancer without smoking causing the cancer. The research was also criticised on the grounds that the answers to questions about past smoking habits might be unreliable, due to either inaccurate memory or deliberate fibbing.

127

In 1951, Doll and Hill set about obtaining data that would be more convincing. Instead of checking the past history of people who already had lung cancer, they decided to select a group of healthy people and monitor who developed the disease. They chose British doctors as their subjects. Perhaps surprisingly, many doctors were smokers at the time. Doll and Hill reckoned that doctors would be more honest and reliable in recording their smoking habits. It was also easy to keep track of doctors because they have to be registered with the General Medical Council. In all, the smoking habits and death rate of over 34 000 doctors were recorded for the next 50 years.

## EXAMPLE

### The evidence that smoking causes disease

**Table 7.1** Some results of the Doll and Hill study of doctors.

| Cause of death | Mortality rate of male doctors /deaths per 1000 per year | | | | |
| | Never smoked | Given up smoking | Still smoking: cigarettes per day | | |
| | | | 1–14 | 15–24 | 25 or more |
| --- | --- | --- | --- | --- | --- |
| Lung cancer | 0.17 | 0.68 | 1.31 | 2.33 | 4.17 |
| Cancer of mouth, throat, gullet | 0.09 | 0.26 | 0.36 | 0.47 | 1.06 |
| Other cancers | 3.34 | 3.72 | 4.21 | 4.67 | 5.38 |
| Other chronic lung disease, e.g. emphysema | 0.11 | 0.64 | 1.04 | 1.41 | 2.61 |
| Heart disease | 6.19 | 7.61 | 9.10 | 10.07 | 11.11 |

Table 7.1 shows some of the results of the Doll and Hill study of doctors. The mortality rate is the average number of deaths per year per 1000 doctors from each cause. A mortality rate of 2.0 means that, on average, two doctors in every thousand died each year. Out of 30 000 doctors, this would mean a total of 60 deaths. However, in this study the rates were calculated separately for each 5 year age group, for example for doctors aged 55–59. The rates were then adjusted because the number of deaths for the younger age groups was much lower than in the older groups. The figures therefore indicate the effect of each cause on both the overall death rate per 1000 doctors and the age of death.

1 What do the data show about the effect of smoking on mortality from lung cancer?
*Mortality is increased significantly by smoking. Even smoking fewer than 15 cigarettes per day increases the risk by as much as seven times.*

2 What do the data show about the effect of smoking on the other causes of death shown in the table?
*Other cancers also increase the mortality rate, especially those associated with the mouth, throat and gullet. The increased risk is much lower for other cancers. Death from other lung diseases is also significantly increased by smoking. There is also a smaller increased risk of heart disease.*

3 For each cause of death, calculate by how much the risk is increased by smoking 25 cigarettes a day compared with not smoking at all.
*Lung cancer, ×24.5; mouth/throat/gullet, ×11.8; other cancers, ×1.6; other lung diseases, ×23.7; heart disease, ×1.8.*

4 Describe the effects of giving up smoking.
*Giving up smoking decreases of lung cancer and associated cancers as well as other lung diseases. It also reduces the chances of dying from other cancers or heart disease.*

5 Describe and explain the results for heart disease.
*There is higher risk of heart disease because several other lifestyle factors contribute to heart disease, such as diet and lack of exercise. By comparison the risk of lung cancer and other lung diseases is low for non-smokers. However non-smokers are liable to die of other diseases, so the death rate is higher, but still somewhat lower than for smokers.*

## ACTIVITY

### Analysing the evidence of the study by Doll and Hill

The graph in Figure 7.24 shows one set of results from this investigation. It shows the percentage of smoking and non-smoking doctors who survived to various ages over the 50 years of the study, which ended in 2001. The smoking group was all those doctors who continued to smoke throughout the investigation, whereas the non-smoking group were those who had never smoked.

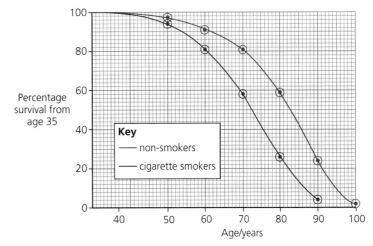

**Figure 7.24** Graph showing the percentage of smoking and non-smoking doctors surviving at each decade of age until age 100.

1 Look at the curve in Figure 7.22 for the number of cigarettes smoked by women. Explain how you would expect the data for deaths from lung cancer in women to give evidence supporting the hypothesis that smoking is a cause of lung cancer.

2 As well as deaths from lung cancer, Figure 7.23 shows the number of deaths from other cancers. How do these data support the hypothesis that smoking causes lung cancer?

3 a) Use the graph in Figure 7.24 to determine the percentage of smokers and non-smokers surviving at ages 60, 70 and 80.

 b) For each of these ages, calculate the difference in the percentage surviving.

 c) By approximately how many years was the average lifespan (length of life) reduced in smokers?

## Gas exchange in plants

The main function of a leaf is to carry out photosynthesis, and for this it needs a supply of carbon dioxide and water. Also, the chloroplasts require good access to sunlight. Sunlight does not penetrate far, and the leaves are thin. In thin leaves there is a short diffusion pathway from the stomata, and there is also a large surface area because of the air spaces surrounding the cells. However, these properties make them vulnerable to damage and dehydration. The leaves of different species are adapted to meet these requirements in a variety of different ways, depending on the conditions in which they live.

## How are leaves adapted for gas exchange?

Figure 7.25 shows the cells and tissues in a section of the leaf of a plant that lives in the relatively moist and cool climate of Britain. Notice how much space there is between the cells in the mesophyll.

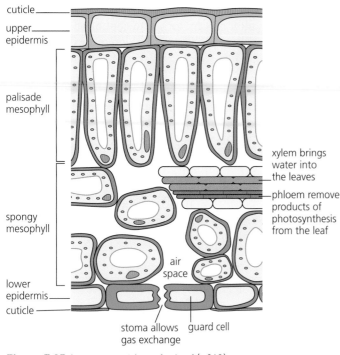

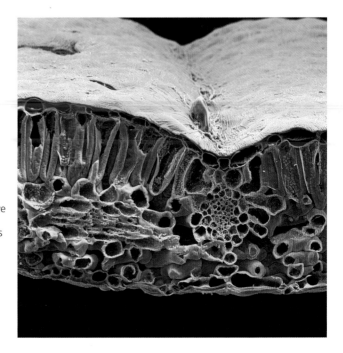

cuticle
upper epidermis
palisade mesophyll
spongy mesophyll
lower epidermis
cuticle
air space
stoma allows gas exchange
guard cell
xylem brings water into the leaves
phloem remove products of photosynthesis from the leaf

**Figure 7.25** A cross-section of a leaf (×210).

The outer cell layer on the upper and lower surface of a leaf is called the **epidermis**. It consists of cells that fit closely together. The outer walls of these cells contain a mixture of a lipid polymer and waxes that make a waterproof **cuticle**. Even in Britain, conditions are usually dry enough for dehydration to be a problem. However, as well as stopping water from escaping, the cuticle prevents most gas exchange. To allow carbon dioxide to get to the photosynthetic cells inside the leaf, there are pores called **stomata** (singular: stoma) in the epidermis. In most leaves, the stomata are mainly on the under-surface of the leaf.

Each stoma is surrounded by a pair of **guard cells**. These are banana-shaped. If the guard cells lose water they become less firm and change shape, closing the stoma, which helps to prevent further water loss. Guard cells are shown in Figure 7.26.

The central tissue of the leaf, called the **mesophyll**, has an extensive network of air spaces, as you can see in Figure 7.25. These spaces allow gases to move to and from the cells by diffusion in the gas phase, which is more rapid than in the liquid phase. There is no active system of ventilation: the thinness of the tissue and distribution of stomata helps to keep the diffusion pathway short. The upper layer, or sometimes two or three layers, is the **palisade mesophyll**. It has elongated cells that contain large numbers of chloroplasts. The **spongy mesophyll**, below it, has more air spaces and the cells have fewer chloroplasts. Water reaches the leaves through the xylem, and we will look in more detail at this process in the Chapter 10.

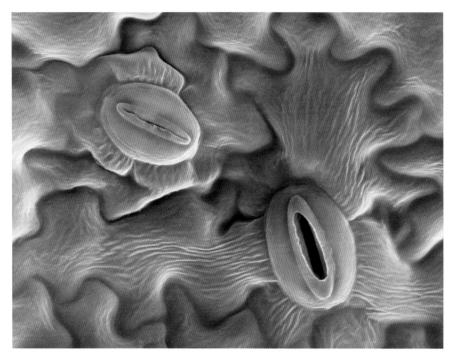

**Figure 7.26** The surface view of guard cells of stomata: closed at top left; open at bottom right.

Despite the amount of fossil fuels that we have burned, the proportion of carbon dioxide in the atmosphere is still low, only about 0.036%. In other words, out of every hundred thousand molecules in the air, fewer than 40 are molecules of carbon dioxide. This means that efficient gas exchange is vital for leaves to get enough carbon dioxide for rapid photosynthesis.

Large numbers of stomata dot the lower surface of leaves. The leaves heat up in sunlight and air that has entered the air spaces warms and rises, so becoming trapped there. This humid air has a high water potential, helping to reduce water loss. The air spaces in the mesophyll allow fast carbon dioxide diffusion to the cells. The palisade cells are elongated, so they have a large surface area exposed to the atmosphere inside the leaf. The large number of chloroplasts in the thin layer of cytoplasm next to the cell wall means that carbon dioxide rapidly takes part in photosynthesis and is therefore removed. This maintains a steep diffusion gradient.

The downside is that the adaptations that boost the entry of carbon dioxide also allow water to be lost rapidly. Water evaporates from the mesophyll cells to become water vapour in the air spaces and, although the cuticle and small pores help to prevent water loss, water vapour can diffuse through the stomata down the water potential gradient. In a damp climate, such as Britain's, plants can also overcome this problem of water loss simply by taking in large amounts of water from the soil. On a hot day, an oak tree may absorb over half a tonne of water, and most of this will evaporate from its leaves. During water shortage, the stomata close to reduce evaporation. Although this tends to stop photosynthesis, it avoids the more catastrophic results of dehydration.

## Counting stomata

A student gathered a sample of leaves from a holly bush. She painted a small area (about the size of a stamp) with clear nail polish on the underside of one leaf. She let this dry. Next she pressed a piece of clear sellotape over the nail-polished area and then carefully peeled it off. She placed this on a clean microscope slide and viewed it under the microscope. She counted the number of stomata visible in the field of view. She repeated this several times. Then she repeated this several times for the top surface of the leaf.

Table 7.2 shows the student's data.

**Table 7.2** Student data.

| Leaf number | Underside of leaf | | Top surface of leaf | |
|---|---|---|---|---|
| | Number of stomata in field of view | Number of stomata /mm$^{-2}$ | Number of stomata in field of view | Number of stomata/mm$^{-2}$ |
| 1 | 41 | 270 | 0 | 0 |
| 2 | 55 | 362 | 0 | 0 |
| 3 | 72 | 474 | 0 | 0 |
| 4 | 52 | 342 | 0 | 0 |
| 5 | 40 | 263 | 0 | 0 |
| 6 | 47 | 309 | 0 | 0 |
| 7 | 38 | 259 | 0 | 0 |
| 8 | 59 | 388 | 0 | 0 |
| 9 | 40 | 263 | 0 | 0 |
| 10 | 42 | 276 | 0 | 0 |
| Mean | | | | |
| SD | | | | |

1 Complete the table to find the mean and standard deviation of these data.
2 Explain the advantage of a leaf having no stomata on the upper surface of the leaves.
3 Suggest how the student worked out the number of stomata per square mm of leaf.
4 Explain the advantage of calculating stomatal density per square millimetre of leaf.
5 Suggest how the student could modify this investigation to find out whether stomatal density is influenced by carbon dioxide concentration.

**TIP**
Find out how to calculate standard deviation in Chapter 14.

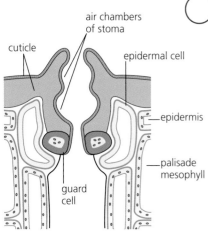

cuticle

air chambers
of stoma

epidermal cell

epidermis

palisade
mesophyll

guard
cell

**Figure 7.27** Sunken stoma.

# How do plants survive in dry environments, such as deserts?

Plants that live in conditions where there is a shortage of fresh water have additional adaptations that enable them to conserve water very effectively. Such plants are called **xerophytes**.

Features that are common in plants growing in environments where there is a shortage of fresh water include:

- thick cuticle
- small or needle-shaped leaves
- few stomata
- stomata sunk into pits in the epidermis (Figure 7.27)
- hairs around the stomata and over the leaf surfaces.

## Deserts

Some of the best-known xerophytes are the cacti, which typically populate the deserts of the western United States. These plants are adapted to desert life by having leaves that are reduced to spines. The leaves have also lost the ability to photosynthesise. Instead, the stem has chloroplasts, as you can see from their green colour. The stem has a large diameter, enabling it to store water in its tissues, and it has a thick cuticle. The number of stomata is much reduced. Overall, the cactus has a low surface area to volume ratio, which reduces water loss. The spines may be just as important as deterrents to grazing animals as in reducing the surface area for evaporation. Figure 7.28 shows the features of a cactus that help it to survive in a dry habitat.

> **TIP**
> Xerophytic plants live in a range of dry environments. Deserts and sand dunes are just two examples of such environments but are not required learning.

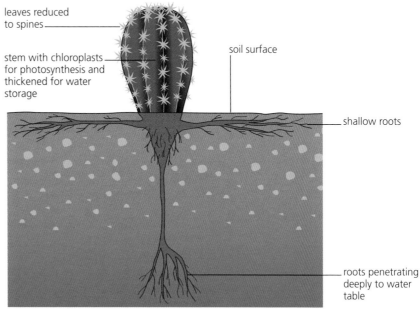

leaves reduced
to spines

stem with chloroplasts
for photosynthesis and
thickened for water
storage

soil surface

shallow roots

roots penetrating
deeply to water
table

**Figure 7.28** Features of a cactus that help it to survive in a dry habitat.

> **TIP**
> You do not need to know specific examples of xerophytes. Make sure you are able to explain the adaptations of any example of xerophyte though.

## Sand dunes

Deserts are not the only habitats that require plants to be adapted to dry conditions in order to survive. The water supply in sand dunes is also limited. Marram grass is an example of a plant that is particularly well adapted to life in dry conditions, and as a result it is one of the first species to colonise new sand dunes on a beach.

133

Leaves of marram grass have several adaptations that help to reduce water loss. In Figure 7.29, you can see that:

● the leaf can roll up so that only one surface is exposed to the wind
● the exposed surface has a thick cuticle and no stomata
● when the leaf is rolled up the stomata are protected in deep grooves
● the inner surface has many hairs.

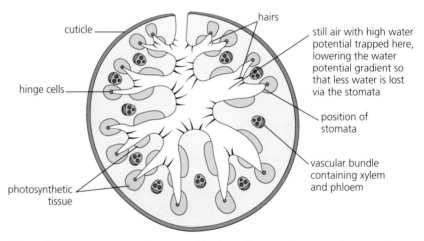

**Figure 7.29** Marram grass is very well adapted to restrict water loss, and is an early coloniser of newly formed sand dunes in the UK.

The advantage of the leaf rolling up is that water vapour becomes trapped and kept near the stomata because there is no wind or air current to move the vapour away. The hairs also help to reduce air movement. The air just outside the stomata becomes much more humid, so it has a higher water potential and the water potential gradient and diffusion of water vapour across the stomata is much reduced. The rate of evaporation is therefore much slower. How does the leaf roll up? In dry conditions, the special hinge cells at the base of the grooves lose water rapidly by evaporation. These cells shrink and pull the sides of the grooves together, making the leaf curl up.

## Factors affecting water loss

The graph in Figure 7.30 shows the difference between the rate of water loss from a leaf without hairs on it in still and in moving air.

1 What does Figure 7.30 show about the effect of air movement on the rate of water loss from a leaf?
   *Air movement replaces humid air with dry air, thus increasing the water diffusion gradient out of the plant, and so increasing water loss.*

Table 7.3 gives the results of measuring the rate of water loss in leaves from three species of flowering plant with differing amounts of hairiness on the leaves.

**Table 7.3** Rate of water loss for flowering plants with leaves of different hairiness.

| Name of flowering plant | Hairiness of leaves | Rate of water loss from leaf surface/$g\,cm^{-2}\,h^{-1}$ |
|---|---|---|
| Sweet violet | Slightly hairy | 0.04 |
| Storksbill | Quite hairy | 0.09 |
| Woundwort | Densely hairy | 0.13 |

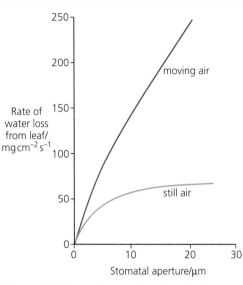

**Figure 7.30** Rate of water loss for a leaf without hairs on it in still air and in moving air.

2 What do these results tell you about the effect of hairs on the rate of water loss in these particular plants?
   *You would expect from the data in the graph of Figure 7.30 that in Table 7.3 the rate of water loss from the hairiest leaves would be lowest because they would trap a layer of still air best. However, the results in the table do not fit with this hypothesis. They clearly do not suggest that the hairs on the woundwort leaves are adaptations to reduce water loss. They may have some other function, such as deterring animals from eating them. Alternatively, any reduction in water loss may be offset by different numbers of stomata, or by some other factor.*

**TIP**
Look carefully at data before assuming that the answer you are expecting is correct!

**TEST YOURSELF**
20 Marram grass grows in sand dunes, close to the sea. Explain the advantages of its xerophytic adaptations.
21 Many cacti have stomata that only open at night. This means they take in carbon dioxide at night, and store it for use in photosynthesis by day. Explain the advantage of this.
22 Make a list of the features of gas exchange in a leaf, using the headings 'large surface area', 'short diffusion distance' and 'large concentration gradient'.
23 Deciduous trees that lose their leaves in winter tend to have large, thin and flat leaves. Explain the advantage of this.
24 Coniferous trees, that do not lose their leaves in winter, tend to have needle-like leaves. Explain the advantage of this.

# Practice questions

1 Describe and explain three features that reduce transpiration in a xerophytic plant. (3)

2 **a)** Insect spiracles often have valves that enable them to be closed, and are often surrounded by tiny hairs. Suggest the advantage of these. (2)

**b)** Describe how oxygen reaches a respiring muscle cell in an insect. (3)

3 **a)** The figure shows *Paramecium*, a single-celled organism.

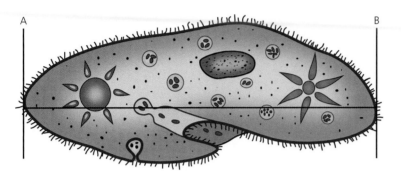

**i)** The actual length of this cell from A to B is 240 μm. Use this information to calculate the magnification of the diagram. Show your working. (2)

**ii)** *Paramecium* does not have a gas exchange system. Explain how gas exchange occurs in this organism. (2)

**b)** Explain how the counter-current principle is important in gas exchange in fish. (2)

4 The table shows some data relating to three different species of fish.

| Species | A | B | C |
|---|---|---|---|
| Swimming speed | Fast | Slow | Slow |
| Water or air breathing | Water | Water | Air |
| Surface area of gills /cm² g⁻¹ | 13.5 | 1.9 | 0.6 |

**a)** Suggest why the surface area of the gills is given per gram of body mass. (2)

**b)** Use the information in the table to explain the advantages of the following features:

**i)** the relative surface area of the gills of species A is larger than the relative surface area of the gills of species B. (2)

**ii)** the relative surface area of the gills of species B is larger than the relative surface area of the gills of species C. (4)

**5** The figure shows part of the tracheal system in an insect that carries out gas exchange.

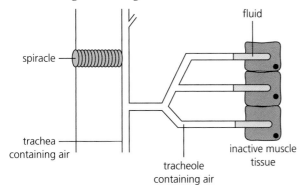

a) **i)** Describe how oxygen in a tracheole reaches a respiring muscle cell in an insect. (2)

**ii)** When insect muscle respires rapidly, products of respiration accumulate in the cells, lowering their water potential. Suggest how this increases the supply of oxygen to active muscle. (3)

b) The figure shows the number of times the spiracles of a flea open when the flea is exposed to various concentrations of oxygen.

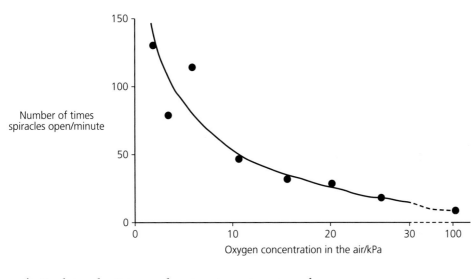

**i)** Explain why it is an advantage to an insect, such as a flea, to be able to close its spiracles. (1)

**ii)** Describe and explain the results shown in the graph. (3)

**6** In the early stages of an asthma attack a man breathed in as deeply as he could. He then breathed out as fast and forcefully as possible through a machine that measured the volume of air as he breathed out. The man then used his inhaler. The inhaler contains a drug that makes muscles relax. Twenty minutes later he did the same test again. The graph overleaf shows the results of the two tests.

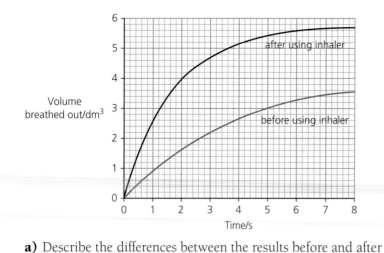

a) Describe the differences between the results before and after using the inhaler. (2)

b) i) The maximum volume of air forcibly breathed out in 1 second is called the FEV1. Use the graph to calculate the FEV1 before and after using the inhaler. (2)

ii) Describe how you could find the initial rate at which air is expelled from the lungs. (2)

c) Describe what the graph shows about the effect of using the inhaler. (2)

d) Explain why the total volume of air breathed out before using the inhaler was lower than the volume expired afterwards. (1)

e) Suggest how using the inhaler caused the difference. (2)

7 a) Use the following information to calculate the approximate surface area of the alveoli when the lungs are inflated. (2)

Each lung has about 350 million alveoli. The mean diameter of each alveolus is about 0.20 mm. The alveoli are not complete spheres because each has an opening. Assume that the surface area is reduced by about 20% compared with a complete sphere. The formula for calculating the surface area of a sphere is $4\pi r^2$.

b) Suggest why it is unlikely that the whole of this area would be available for diffusion of oxygen. (1)

## Stretch and challenge

8 People who have had a spinal cord injury, resulting in paralysis, might need to use a ventilator. Explain the reason for this and how a ventilator works.

9 In the first half of the twentieth century, poliomyelitis was a fairly common disease in the UK. Children who had polio were sometimes so paralysed that they had to be nursed in an iron lung. Explain how an iron lung works. How does an iron lung differ from a ventilator, and what are the advantages of using a ventilator?

10 How does gas exchange occur in molluscs and amphibians? Explain how the gas exchange systems in the two groups of animals are adaptations to their environment.

# 8 Digestion and absorption

# Introduction

When your digestive system becomes infected it is not very pleasant. You may have had diarrhoea. After a day or two, however, it is more than likely that you were over the worst of it. By contrast, every year nearly 2 million children die in developing countries when they lose too much fluid and become dehydrated because of diarrhoeal disease. Tragically, we could prevent many of these deaths with a simple mixture of water, glucose and salts. This mixture is called an oral rehydration solution (ORS).

**Figure 8.1** In overcrowded refugee camps, the water is often contaminated, and it is difficult to maintain good hygiene. As a consequence, people often die of cholera and other diarrhoeal diseases. During the 1971 war for independence in East Pakistan (now Bangladesh), up to 30% of the refugees in the refugee camps in India died when doctors ran out of medicines. In camps where doctors used oral rehydration solutions, the death rate was only 3%.

# What does the digestive system do?

The digestive system consists of the gut, which forms a tube extending from the mouth at one end, through the body, to the anus at the other end. Food is ingested: it is taken in. In the mouth, it is chewed, mixed with saliva, and swallowed. The food is now inside the gut but it is not yet inside the body itself (to truly enter the body, substances have to cross cell-surface membranes). Before it can pass through the wall of the gut and into the blood, it must be digested. The food is mixed with digestive juices secreted by various glands as it is squeezed and pushed along by the muscular walls of the gut.

Figure 8.2 shows the human digestive system. The stomach and the first part of the small intestine are where food is digested. The digestive juices, made by gland cells of the digestive system, contain enzymes. The enzymes act on the large insoluble molecules of protein, starch and fats that are the main components of our food. The enzymes hydrolyse them into smaller soluble molecules that can be transported across cell-surface membranes:

- protein is hydrolysed to amino acids
- starch is hydrolysed to simple sugars
- fats are hydrolysed to a mixture of fatty acids and glycerol.

These small molecules are absorbed through the lining of the small intestine into the blood and transported to the body's cells.

**1. Ingestion**
Food is taken into the mouth.

**2. Digestion**
Enzymes break large insoluble molecules into smaller soluble ones.

**3. Absorption**
The products of digestion are absorbed through the lining of the intestine.

**4. Egestion**
Removal of faeces containing:
- undigested food
- bacteria
- cells from the intestine lining
- enzymes.

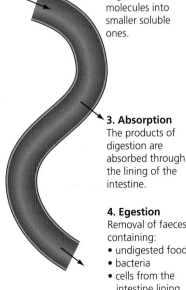

**Figure 8.3** Processing food.

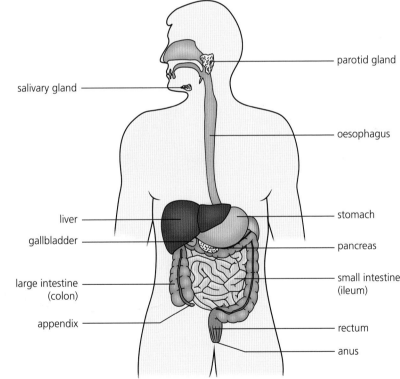

parotid gland

salivary gland

oesophagus

liver

stomach

gallbladder

pancreas

large intestine (colon)

small intestine (ileum)

appendix

rectum

anus

**Figure 8.2** The human digestive system.

Some substances in food, such as cellulose, cannot be digested by the human gut. They pass out through the anus, together with cells scraped from the gut lining, enzymes and bacteria, and are egested as faeces. Figure 8.3 summarises these processes.

# Digesting carbohydrates

**Figure 8.4** We eat different carbohydrates in our diet. Bread, pasta and potatoes all contain starch; fruit contains sucrose, glucose and fructose. Milk contains lactose.

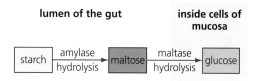

**Figure 8.5** Disaccharidases such as maltase are in the cell-surface membrane of epithelial cells in the small intestine.

Respiration is the biochemical process by which we produce ATP from energy stores such as glucose. The glucose we use in respiration comes mainly from the carbohydrates that we eat and derivatives of lipids and proteins (Figure 8.4).

Any glucose that we eat can be absorbed as soon as it reaches the small intestine because glucose molecules are small and are able to be carried through the cell-surface membranes of the cells that line the intestine and enter the blood. Starch, though, has large insoluble molecules that cannot be absorbed. They must be digested, first to maltose and then to glucose. This involves the enzymes amylase and maltase.

Carbohydrate digestion occurs in the mouth and small intestine. Amylase is secreted in saliva and starts to digest starch until it is denatured by the acidity of the stomach. Amylase is secreted into the small intestine in pancreatic juice from the pancreas. Amylase hydrolyses starch to maltose (a disaccharide). The cells lining the small intestine have the enzyme maltase in their cell-surface membranes. Maltase hydrolyses maltose to glucose (Figure 8.5).

The glucose is released into the epithelial cell. This is shown in Figure 8.6. There are other disaccharidases in the membranes of the epithelial cells in the small intestine. These include sucrase and lactase (in some individuals).

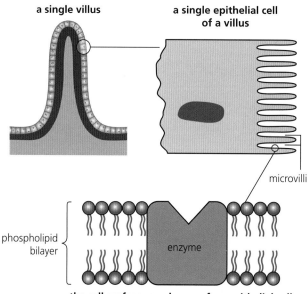

**Figure 8.6** The lining of the small intestine, showing a villus, microvilli on the surface of an epithelial cell of a villus and its cell-surface membrane.

The lining of the small intestine is folded into finger-like projections called villi. These increase the surface area for absorption. The epithelium of the villi is made of epithelial cells that have microvilli on their surface (see Chapter 3). There are enzymes in the cell-surface membrane of these epithelial cells. Figure 8.7 summarises carbohydrate digestion.

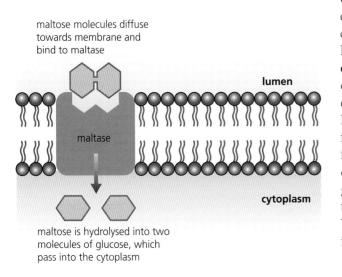

maltose molecules diffuse towards membrane and bind to maltase

lumen

maltase

cytoplasm

maltose is hydrolysed into two molecules of glucose, which pass into the cytoplasm

**Figure 8.7** A summary of carbohydrate digestion in humans.

One of the most important products of carbohydrate digestion is glucose. The intestine absorbs glucose by a combination of facilitated diffusion and active transport. Look at Figure 8.8. There are carrier molecules called **co-transport proteins** in the cell-surface membranes of the epithelial cells. These are carrier molecules that only transport glucose in the presence of sodium ions. Each time a sodium ion is transported into the cell, so is a glucose molecule. This part of the process involves facilitated diffusion. Facilitated diffusion, however, will only work if substances can move down a concentration gradient. This gradient is maintained by actively transporting sodium ions out of the cell into the blood. The glucose molecules pass from the inside of the cells into the blood by facilitated diffusion.

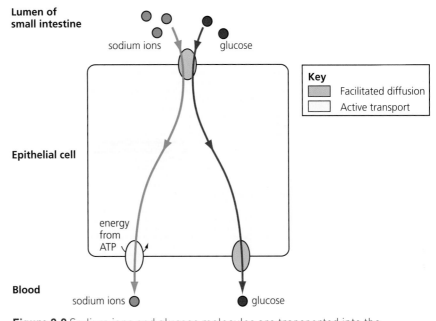

**Figure 8.8** Sodium ions and glucose molecules are transported into the epithelial cells lining the small intestine by facilitated diffusion. Sodium and glucose enter the cells through co-transport proteins. The sodium ions are then actively transported from the epithelial cells and into the blood.

## TEST YOURSELF

1 The epithelial cells in the small intestine contain many mitochondria. Explain the link between the large number of mitochondria and the transport of sodium ions out of the cells into the blood.

2 Explain why cellulose cannot be digested in the human gut.

3 Draw a molecule of maltose and show how it is hydrolysed into two molecules of glucose.

4 Explain why the carrier protein in Figure 8.8 carries glucose across the membrane but not other molecules such as fructose.

## EXAMPLE

### Investigating absorption

Scientists investigated how different factors affected the rate of absorption of glucose from a piece of small intestine. The results of their investigation are shown in the graph in Figure 8.9.

1 Look carefully at the graph. What are the independent variables in this investigation?
*There are two: the concentration of glucose in the lumen of the intestine, and whether or not the glucose solution was stirred. They are the independent variables because they were the factors that the scientists changed.*

2 What is the dependent variable?
*The dependent variable is the factor that is measured as a result of changing the independent variable. In this investigation, it is the rate of absorption of glucose from the small intestine.*

In this investigation, the scientists measured the rate of glucose absorption.

3 What is meant by the rate of absorption?
*The rate is the amount absorbed divided by the time taken. We often use rates in biology because they allow comparisons to be made.*

When the scientists carried out the investigation, they kept the temperature constant.

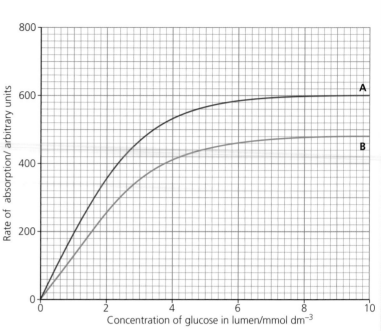

**Figure 8.9** The graph shows the effect of glucose concentration on the rate of absorption of glucose from the small intestine. Curve A shows the results when the glucose solution in the intestine was stirred. Curve B shows the results when the glucose solution was not stirred.

## TIP
Look at investigating scientific questions in Chapter 15, Practical skills.

4 Why did the scientists keep the temperature constant?
*Temperature affects the rate of absorption. If the scientists allowed the temperature to vary as well as, say, the concentration of glucose in the lumen of the intestine, they would not know what caused any change in the rate of absorption.*

5 Now look at curve B on the graph. Describe how the concentration of the glucose solution in the lumen of the small intestine affects the rate of absorption.
*The rate of absorption increases as the concentration of glucose in the small intestine increases, and then gradually levels off. After a concentration of approximately 5 mmol dm$^{-3}$, it remains constant.*

6 The rate of absorption is more or less constant above a concentration of 5 mmol dm$^{-3}$. Explain why.
*This is another graph where limiting factors are involved (see pages 31–32 where we looked at the effect of substrate concentration on the rate of an enzyme-controlled reaction).*

There must be something other than the concentration of glucose in the lumen of the small intestine that is limiting the rate of absorption here. It is probably the number of glucose carrier molecules in the cell-surface membrane of epithelial cells lining the intestine.

7 Describe and explain the effect of stirring on the rate of absorption.
*The graph shows that stirring increases the rate of absorption, regardless of the concentration of glucose in the lumen of the small intestine. Think about what happens when the glucose solution has not been stirred. As it is absorbed into the cells, the concentration in the intestine will fall, and the difference in lumen and cell concentrations will become less and less. Obviously, this fall in the concentration gradient will slow the rate of diffusion. Stirring maintains the concentration gradient. This results in a higher rate of absorption.*

## Starch and colon cancer

Scientists investigated the relationship between the food we eat and the **probability** of developing cancer of the colon. The colon is the last part of the digestive system. One of the factors that the scientists looked at was the amount of starch in people's diet. The scattergram in Figure 8.10 shows some of their results.

**Probability** A mathematical way of expressing the likelihood of a particular event occurring. You could describe the likelihood of a person developing colon cancer as 1 in 1000 (or 0.001) so you should use the term probability. In short, if you could put a number to it, use probability.

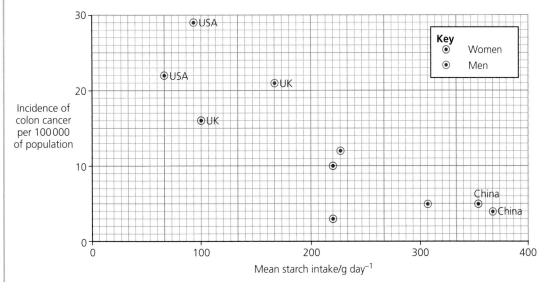

**Figure 8.10** A scattergram showing the incidence of colon cancer plotted against the mean amount of starch in the diet for men and women of different nationalities.

> **TIP**
> In considering any set of data, one of the first things we do is to look carefully at the data set and make sure that we understand exactly what it shows. We will start here by looking at the axes.

1 The *y*-axis shows the incidence (number of cases) of colon cancer per 100 000 of the population. Why are the figures given per 100 000 of the population?
*This is a straightforward question to answer. We want to compare the number of cases of cancer in different groups of people. The only way to do this is to compare like with like. The population of China is around 1 billion. The population of the UK is only about 65 million. In view of this, it is very likely that China will have more cases of colon cancer, simply because more people live there. Looking at the incidence per 100 000 allows us to make a fair comparison.*

2 Do you think that giving the starch intake in grams per day lets us make a fair comparison?
*It certainly helps because we must make sure that in each case we compare the amount of starch eaten over the same period of time. But people also vary in size. American men, for example, are larger on average than Chinese men. This probably affects the amount they eat. It might have been better to have taken body size into account as well, in which case the figures for starch intake would be grams per day per kilogram of body mass ($g\,day^{-1}\,kg^{-1}$).*

3 Why did the scientists plot the figures for men and for women separately?

*There are several possible reasons for this, but what they all come down to is that men and women form separate groups. They differ in body size and so probably eat different amounts of starch. There are also other important differences. Women may become pregnant, and they have different concentrations of different hormones circulating in their blood. These are factors that could affect the probability of developing colon cancer. But the scientists did not collect data about these factors. So it is better to treat men and women as separate groups.*

4 Is there a correlation between the amount of starch that people eat and the probability of developing colon cancer?

*We can find out whether there is such a correlation in several ways. Figure 8.11 shows that we can do this by drawing the line of best fit on the scattergram.*

*As you can see, the line slopes downwards. It tells us that the more starch people eat, the lower is the probability that they will develop colon cancer. American men eat very little starch. They have the highest incidence of colon cancer. Chinese men, however, eat a lot of starch and they have the lowest incidence of colon cancer.*

**TIP**

Find out more about correlation and using scattergrams in Chapter 14.

5 Does this mean that eating starch lowers the incidence of colon cancer?

*We have to be very careful here. Just because two things are correlated, it doesn't mean that one causes the other. We have seen that there seems to be a clear relationship between the amount of starch in the diet and the incidence of colon cancer, but we cannot say that eating a lot of starch will keep a person free of colon cancer. Other things could be involved. People in the USA probably eat more protein or more fat than those who live in China. Maybe that is the reason for the higher incidence of colon cancer. In other words, there could be other factors involved that we haven't considered.*

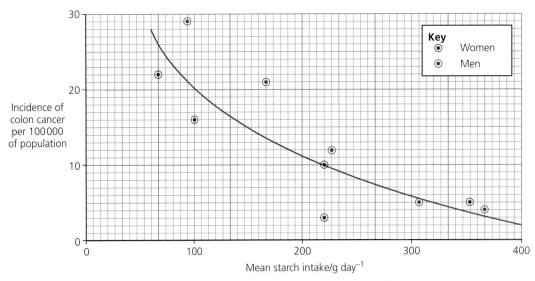

**Figure 8.11** A scattergram showing the incidence of colon cancer plotted against the mean amount of starch in the diet for people of different nationalities. The line of best fit has been added.

6 So, does eating starch lower the incidence of colon cancer?

*This is where scientists use their biological knowledge to suggest possible explanations for the results they collect. Food such as a banana contains different sorts of starch. Some of the starch in banana is digested only slowly in the human intestines. It is called resistant starch. When resistant starch enters the last part of our digestive system, the colon, it is hydrolysed by the bacteria that live there. They produce substances such as butyric acid when they digest starch.*

*Resistant starch may help to prevent cancers developing in one of two ways. First, butyric acid is known to kill cancer cells. Second, resistant starch helps to increase the rate of movement of faeces through the colon. This means that any substances in the faeces that could cause cancer spend less time in contact with the cells that line the colon. Before we can say definitely what happens, a lot more work is necessary. On the evidence that we have here, all we can conclude is that it is possible that eating starch lowers the incidence of colon cancer.*

## Protein digestion

Proteins are digested by enzymes called **proteases**. The process of protein digestion starts in the lumen of the stomach. Here an enzyme called an **endopeptidase** hydrolyses peptide bonds within the protein, hydrolysing it into smaller polypeptide 'chunks'. The endopeptidase in the stomach is secreted with hydrochloric acid, so the pH is very low in the stomach. This was particularly important in early humans as the acidic conditions would have destroyed many of the bacteria and parasites in their food.

After the stomach, the partly digested food passes into the small intestine. Here, pancreatic juice neutralises the acidic mixture that leaves the stomach and contains both endopeptidases and exopeptidases. Exopeptidases hydrolyse near the ends of the polypeptide chains, producing dipeptides. This is shown in Figure 8.12.

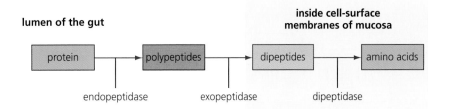

**Figure 8.12** A summary of protein digestion in humans.

Finally there are dipeptidase enzymes in the cell-surface membrane of the epithelial cells of the small intestine (Figure 8.13, overleaf). These hydrolyse dipeptides and release amino acids into the cytoplasm of the cell.

147

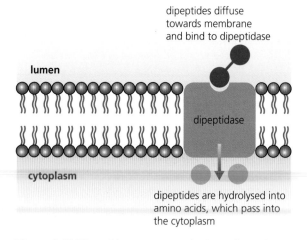

dipeptides diffuse
towards membrane
and bind to dipeptidase

**lumen**

dipeptidase

**cytoplasm**

dipeptides are hydrolysed into
amino acids, which pass into
the cytoplasm

**Figure 8.13** Dipeptidase enzymes in the cell-surface
membrane of the epithelial cells of the small intestine.

In the membrane of the epithelial cells of the small intestine, there are
amino acid carrier proteins, similar to the glucose carrier proteins you
learned about earlier in the chapter (see page 143). These also rely on
sodium ions. The sodium/potassium pump transports sodium ions out
of the cell by active transport. Amino acids and sodium ions bind to the
carrier protein. As the sodium ions diffuse into the cell, amino acids are
carried too. The amino acids then diffuse to the other end of the cell.
They are then transferred to the capillaries by facilitated diffusion. This is
shown in Figure 8.14.

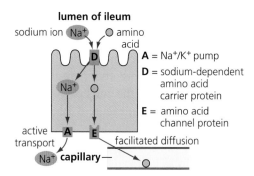

**lumen of ileum**

sodium ion $Na^+$      $\bigcirc$ amino
acid

D

$Na^+$       $\bigcirc$

active
transport

A      E

$Na^+$ **capillary**

**A** = $Na^+$/$K^+$ pump

**D** = sodium-dependent
amino acid
carrier protein

**E** = amino acid
channel protein

facilitated diffusion

**Figure 8.14** Absorption of amino acids
from the lumen of the small intestine
into the blood capillary.

**TEST YOURSELF**

**5** Explain the advantage of endopeptidase enzymes hydrolysing proteins
before exopeptidases.

**6** The lining of the stomach is covered in thick mucus. Explain why.

**7** Use a diagram to show how a dipeptide is hydrolysed into amino acids.

**8** List the similarities and differences between carbohydrate and protein
digestion.

# ACTIVITY

## Chromatography of amino acids

**TIP**

See Chapter 15 for more information on chromatography.

A student wanted to identify the amino acids present in an unknown food substance, she decided to use a technique called paper chromatography to separate the different amino acids. She used the method below.

She drew an origin line in pencil and placed several crosses on the origin line. She placed a concentrated spot of the mixture on the origin line, and then put concentrated spots of known amino acids on the other origins on the paper. She placed the paper in a tank containing a solvent (butanol and ethanoic acid) for 2 hours. There was only a small amount of solvent in the tank, so that the solvent did not go above the origin line on the paper. When the solvent had almost reached the top of the paper, she quickly marked the solvent front using a pencil. She sprayed the chromatogram with ninhydrin and placed the paper in an oven at 100°C. After this the amino acids showed up as blue/purple spots. A drawing of the student's chromatogram is shown in Figure 8.15.

She calculated the $R_f$ value using this formula:

$$R_f = \frac{\text{Distance moved by the solute}}{\text{Distance moved by the sovent}}$$

The $R_f$ value of each solute is calculated and compared to published values in the same solvent. ($R_f$ values are always less than 1 and have no units.)

1 The student wore plastic gloves while handling the chromatography paper. Suggest why.
2 The origin line was drawn in pencil, not ink. Suggest why.
3 Why was it important that the solvent in the tank did not come above the origin line?
4 Ninhydrin is a locating agent. Suggest why it is needed.
5 Calculate the $R_f$ values of the spots on the chromatogram. Use your calculations to identify the amino acids present in the mixture.

| Amino acid | $R_f$ value |
|---|---|
| Alanine | 0.38 |
| Arginine | 0.20 |
| Glutamine | 0.13 |
| Leucine | 0.73 |
| Methionine | 0.55 |
| Tyrosine | 0.45 |

**TIP**

Chromatography is not a required practical until your second year of study.

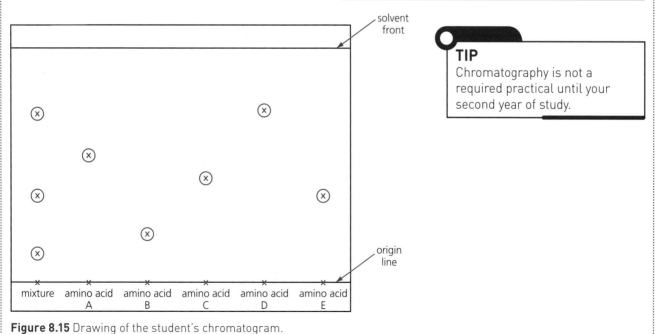

Figure 8.15 Drawing of the student's chromatogram.

# Lipid digestion

Lipid digestion only occurs in the lumen of the small intestine. In the stomach, solid lipids are churned into a fatty liquid made of fat droplets, but no digestion takes place in the stomach. Once the fatty liquid enters the first part of the small intestine, bile from the gall bladder, which is connected to the liver, is secreted. This liquid contains bile salts. Bile salts bind to the fat droplets and break them down into smaller fat droplets. This is called **emulsification**. This is not digestion, but a physical process that increases the surface area available for lipase enzymes to digest the lipids. Lipase secreted into the small intestine by the pancreas hydrolyses lipids into fatty acids and glycerol. This is summarised in Figure 8.16.

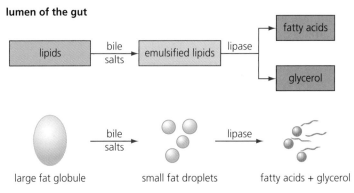

**Figure 8.16** A summary of lipid digestion in humans.

Lipid absorption is different from that of proteins and carbohydrates. Lipids are hydrolysed into glycerol, fatty acids and monoglycerides (partly broken down lipids). Monoglycerides and fatty acids associate with bile salts and phopholipids to form **micelles**. Micelles are droplets that are about 200 times smaller than the small fat droplets in the emulsified lipids. Micelles transport the poorly soluble monoglycerides and fatty acids to the surface of the epithelial cells where they can be absorbed. Micelles are small enough to fit between the microvilli. Micelles constantly break down and re-form, building up a small pool of monoglycerides and fatty acids that are in solution. Only freely dissolved monoglycerides and fatty acids can be absorbed, not the micelles. Because they are non-polar, monoglycerides and fatty acids can diffuse through the phospholipid bilayer of the plasma membrane of the epithelial cell.

Fatty acids can have different lengths of hydrocarbon chains in them. Short-chain fatty acids diffuse directly into the blood from the lumen of the small intestine via the epithelial cells. They can pass easily through the membranes because they can diffuse through the phospholipid bilayer. Longer-chain fatty acids, monoglycerides and glycerol diffuse into the epithelial cells where they recombine to form triglycerides again. These triglycerides are packaged with cholesterol and phospholipids to form water-soluble fat droplets called **chylomicrons**. These are transferred to a lymph vessel inside the villus, called a lacteal, by exocytosis. Exocytosis is when a small piece of the cell-surface membrane is wrapped around the lipid droplets and pinched off, so that the fatty droplets are now wrapped in membrane as they enter the lymph vessels. They eventually enter the blood system. This is shown in Figure 8.17.

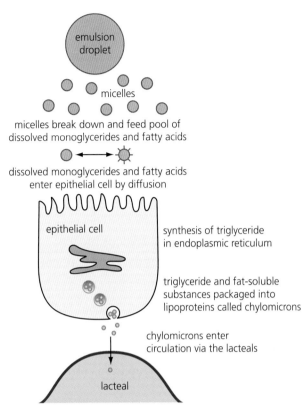

emulsion droplet

micelles

micelles break down and feed pool of dissolved monoglycerides and fatty acids

dissolved monoglycerides and fatty acids enter epithelial cell by diffusion

epithelial cell

synthesis of triglyceride in endoplasmic reticulum

triglyceride and fat-soluble substances packaged into lipoproteins called chylomicrons

chylomicrons enter circulation via the lacteals

lacteal

**Figure 8.17** Absorption of lipids from the lumen of the small intestine to the blood capillaries and the lymph system.

## TEST YOURSELF

**9** Explain why the products of lipid digestion can be absorbed by simple diffusion.

**10** Explain how bile helps in lipid digestion.

**11** How does lipid digestion differ from carbohydrate and protein digestion?

**12** Explain why it is important for chylomicrons to be water-soluble.

# Practice questions

1 **a)** Explain the advantage of an epithelial cell from the small intestine

    **i)** having many microvilli on its surface (3)

    **ii)** containing many mitochondria. (3)

  **b) i)** Describe how a piece of cheese, composed mainly of fat, is digested in the human gut. (3)

    **ii)** Describe how the digestion products of the cheese are absorbed in the gut. (3)

2 In the food industry, enzymes are used to produce glucose and maltose from starch. Food scientists use the expression dextrose equivalent (D.E.) when they are describing the products formed from the starch. D.E. is calculated from the formula:

$$\text{D.E.} = 100 \times \frac{\text{Number of glycosidic bonds broken}}{\text{Initial number of glycosidic bonds present}}$$

  **a) i)** The D.E. of glucose is 100. Explain why. (1)

    **ii)** What would you expect to be the D.E. of maltose formed from starch? Explain how you arrived at your answer. (2)

The flow chart summarises the processes in which enzymes are used to produce glucose syrup and maltose syrup from corn starch.

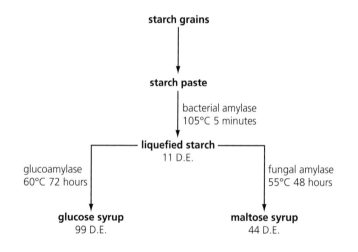

  **b)** Explain the evidence from the flow chart that:

    **i)** bacterial amylase is a thermostable enzyme (2)

    **ii)** the glucose syrup formed in this process is not pure. (2)

  **c)** What does the information in this flow chart suggest about the chemical nature of liquefied starch? Explain your answer. (3)

  **d)** Glucose used to be made by hydrolysing starch with acid. The use of enzymes has a number of advantages over using acid.

    **i)** When the process of converting liquefied starch to glucose is carried out using enzymes, it can be stopped very easily. This is done by heating the mixture to 85°C for 5 minutes. Explain how heating the mixture stops the reaction. (2)

    **ii)** Suggest one other advantage of using enzyme hydrolysis rather than acid hydrolysis. (1)

**3 a)** Lactose is a disaccharide found in milk. Describe how it is digested and absorbed in the human gut. (2)

  **b)** Many adults are lactose intolerant. This means that they no longer produce the enzyme lactase in their small intestines. Two people were tested for lactose intolerance. Both people had nothing to eat, and only water to drink, for 10 hours. Then both people were given a drink containing 50 g of lactose. The concentration of glucose in their blood was measured over the next 2 hours. The results of this test are shown in the graph.

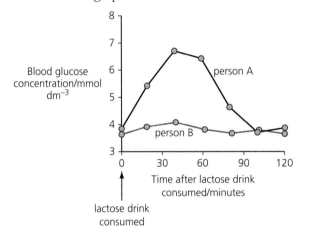

    **i)** The people had nothing to eat, and only water to drink, for 10 hours before the test. Explain why. (1)

    **ii)** Which one of these people was lactose intolerant? Explain the reason for your answer. (2)

  **c)** How could you calculate the percentage increase in blood glucose concentration for person A in the first 30 minutes after consuming the lactose drink? (1)

## Stretch and challenge

  **4** How do animals such as rabbits and cattle survive on a diet consisting mainly of cellulose without producing any cellulase enzymes?

  **5** Compare the gut of a bird with a human gut. Explain how the differences show adaptations to different diets.

# 9 Mass transport in animals

## Introduction

The exchange of substances between the internal and external environments of organisms takes place at **exchange surfaces**. But substances have not really entered or left organisms until they enter or leave their cells by crossing cell-surface membranes. In large multicellular animals, the exchange surfaces you have seen in chapters 7 and 8 are some distance away from most of the rest of their cells. This is also the case for the exchange surfaces of multicellular plants. This means that exchange surfaces such as the lungs and digestive system are linked to mass transport systems that carry substances the relatively large distances to other body cells and from those cells back to the exchange surfaces. Diffusion across such distances

Mass transport The bulk movement of liquids or gases in one direction, usually through a system of tubes or vessels.

would not meet the metabolic demands of cells. But diffusion is still involved. It just takes place at each end of the route that substances travel, at the exchange surfaces and between the cells and their immediate fluid environment. Mass transport carries substances quickly from one to the other but also maintains the diffusion gradients at the exchange surfaces and between the cells and their fluid surroundings. In this way, mass transport helps to keep the immediate fluid environment of cells in multicellular organisms within a suitable metabolic range for effective cell activity.

# Blood and circulation

## Blood

Blood has two main constituents, the liquid plasma and the cells.

When blood is centrifugated, the red blood cells are forced to the bottom of the tube because they are heaviest. Most of the remainder, pale yellow **plasma**, is a fluid that makes up just over half the volume. Its role is to transport dissolved glucose, amino acids, urea, mineral ions and hormones. The great majority of cells are the red blood cells (technically called erythrocytes), which transport oxygen and some carbon dioxide. There are many fewer white cells (leucocytes), and these form a barely visible layer on top of the red cells.

The proportions of plasma and cells are important. There has to be enough plasma for the blood to flow easily. On the other hand, an efficient oxygen supply requires a large proportion of red blood cells. A quite small decrease in the proportion of plasma makes the blood more viscous (sticky), and the heart has to pump harder to push it round. This increases blood pressure and places additional strain on the heart. Increased blood viscosity is one way in which dehydration can affect the performance of an athlete.

The human heart is the pump that forces blood round the body. In effect, it is a double pump, because the left half pumps blood to the majority of the body tissues, while the right half pumps blood through the lungs. In this section we consider how the vessels that carry the blood around the body are adapted for their function.

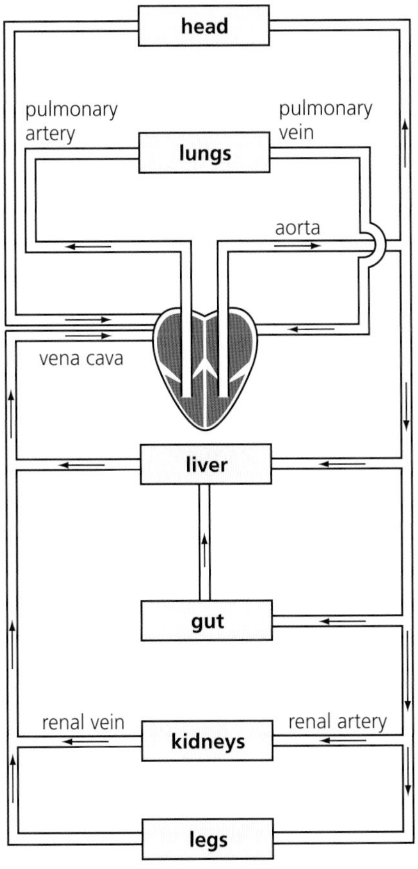

Figure 9.2 is a diagram showing the basic plan of the blood circulation of a mammal. Blood is pumped out of the heart into **arteries**. These branch into narrower **arterioles** within the organs, and the arterioles branch to form the mass of very narrow **capillaries**, which penetrate the tissues. From the capillaries the blood flows into **venules** and **veins** that transport it back to the heart.

## Arteries

The arteries have to withstand the full force of the pumping action of the heart's ventricles. They are adapted for this by having thick but flexible walls. You can see from Figure 9.3 (overleaf) that an artery wall has three layers. The thick middle layer consists of a mixture of muscle cells and elastic fibres. Outside this is a layer of tough protein fibres. The innermost layer, the endothelium, consists of flattened cells with an extremely smooth surface. This reduces friction and ensures that blood flows freely and does not stick to the walls. If this surface does get damaged blood clots are liable to form and may block the artery (see page 174).

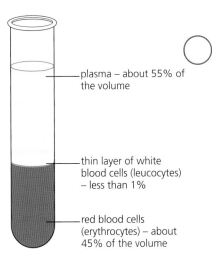

plasma – about 55% of the volume

thin layer of white blood cells (leucocytes) – less than 1%

red blood cells (erythrocytes) – about 45% of the volume

**Figure 9.1** A tube of centrifuged blood.

**Plasma** The liquid component of blood.

head

pulmonary artery

pulmonary vein

lungs

aorta

vena cava

liver

gut

renal vein

renal artery

kidneys

legs

**Figure 9.2** The basic plan of the blood circulation of a mammal.

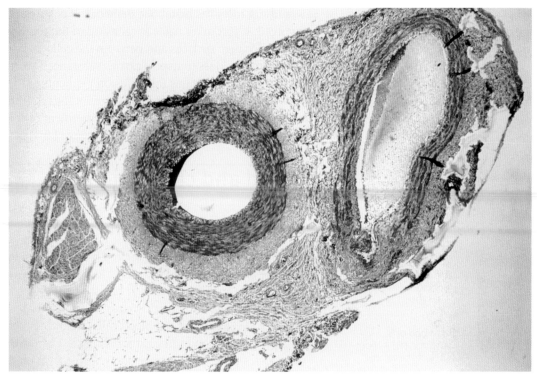

**Figure 9.3** Photomicrograph of the cross-section of an artery (left) and a vein (right).

> **TIP**
> Arteries and veins are defined by the direction of blood flow. Arteries carry blood away from the heart while veins carry blood back to the heart. You should not define them in terms of carrying oxygenated or oxygen-depleted blood because there are exceptions such as the pulmonary artery and the umbilical artery, which both carry oxygen-depleted blood.

The elastic fibres in the middle layer of the artery wall allow the artery to expand each time that the heart beats. This is better than having a rigid tube because the stretching of the fibres absorbs the shock waves caused by the heart's forceful pumping action. The other advantage of the elasticity is that the fibres recoil to their original length between heartbeats. This smoothes out the changes in pressure and maintains a more constant blood flow. Close to the heart, arteries have a high proportion of elastic fibres and few muscle cells. In the smaller arterioles the balance is the opposite way round. Here, muscle cells can contract and partially shut off blood flow to particular organs. For example, during exercise, blood flow to the stomach and intestines is reduced, allowing greater blood flow to the muscles.

## Veins

Veins have much thinner walls than arteries. The blood pressure is much lower and the blood moves more slowly. They have a considerably wider central lumen and so the same volume of blood returns to the heart as leaves it, but the returning flow is slower. It is like a wide river moving slowly compared with a narrow mountain torrent that joins the river. Like the arteries, the veins have walls of three layers, but the middle layer is much less muscular and has only a few elastic fibres. Valves in the veins

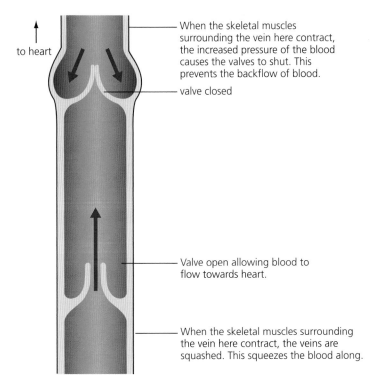

When the skeletal muscles surrounding the vein here contract, the increased pressure of the blood causes the valves to shut. This prevents the backflow of blood.

valve closed

to heart

Valve open allowing blood to flow towards heart.

When the skeletal muscles surrounding the vein here contract, the veins are squashed. This squeezes the blood along.

**Figure 9.4** Diagram of valves in a vein.

ensure that the blood can only flow towards the heart (Figure 9.4). Finally, blood is drawn into the heart when the chambers expand and there is a brief period of lower pressure (see Figure 9.32, page 169).

The smallest arterioles and venules are connected by capillaries. A capillary is only about 8 μm in diameter, and it is estimated that an adult human has nearly 100 000 km of capillaries. That is more than twice the distance round the equator! All of the capillaries put together provide an enormous surface area for the exchange of gases, glucose and other substances, and no cell in the body is more than a very short diffusion distance from a capillary. The walls of many capillaries, especially those in muscles and the lungs, consist of a single thin endothelial cell wrapped into a tubular shape, as shown in Figure 9.5. Others have two or three cells linked together, but all have very thin walls only one cell thick, which allows for rapid diffusion but also increases resistance to flow.

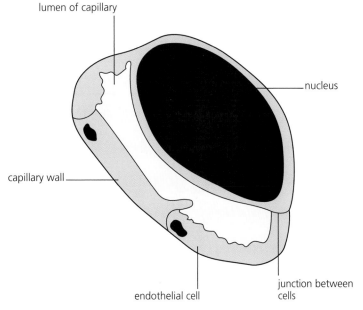

lumen of capillary

nucleus

capillary wall

endothelial cell

junction between cells

**Figure 9.5** A section across a capillary.

**TEST YOURSELF**

1 Give the name of the arteries that supply blood to the kidneys.
2 Explain the advantage of having a lot of elastic tissue in the wall of an artery.
3 Explain the role of the muscle tissue in the wall of an arteriole.
4 Explain the function of valves in veins.

# Leaky vessels

Capillaries have such thin walls that most substances can easily pass through. Only substances with particularly large molecules, such as most proteins, cannot escape from the blood. This has the advantage that oxygen, carbon dioxide, glucose, amino acids, hormones and many other substances are easily exchanged between the blood and the tissues.

The capillaries also allow some fluid to leak out between the cells. This produces a solution that fills the spaces between cells. This is called tissue fluid. Obviously, water cannot simply drain out of the blood vessels without being replaced. In fact, in the tissues there is constant movement of water into and out of the capillaries. By the time blood enters the capillaries from the arterioles it is at a much lower pressure than when it left the heart in the arteries. Nevertheless, this hydrostatic pressure is still large enough to force water out of capillaries into the tissue fluid. Therefore, as the blood passes along the capillaries, the water content decreases. Small solutes including glucose and ions also pass out, but the larger soluble proteins in the blood plasma cannot go through the walls of the capillaries. As water is lost, the concentration of proteins in the blood plasma increases and the pressure drops. This lowers the **water potential** inside the capillary. The result is that at the venule end of the capillary the water potential of the plasma is lower than the water potential of the tissue fluid. Therefore, water goes back into the capillary by osmosis. This is summarised in Figure 9.6.

**Tissue fluid** A fluid surrounding cells that is formed from blood plasma without large proteins. It is the immediate environment of each cell.

**Hydrostatic pressure** Pressure caused by an increased volume of fluid inside a vessel.

> **TIP**
> Refresh your memory of water potential by going back and reading about it in Chapter 3.

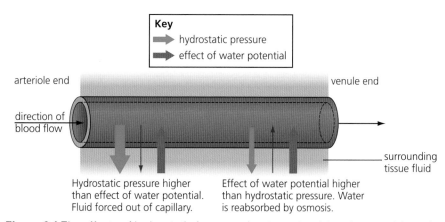

**Figure 9.6** The effects of hydrostatic forces and water potential at the arteriole end and the venule end of a capillary.

Not all of the excess water in the tissues is reabsorbed into the blood capillaries by osmosis. The tissues also have a drainage system that consists of tubes slightly wider than capillaries. These tubes are called

**Lymph capillary** A vessel that helps to drain tissue fluid and return plasma proteins to the blood via the lymphatic system.

lymph capillaries (Figure 9.7). They are not part of a circulatory system like the blood. They have closed ends, which are sufficiently porous for tissue fluid and large molecules to enter. This is important because some **plasma proteins** do escape from the blood capillaries. If they accumulated in the tissue fluid, they would lower the water potential of the tissue fluid until water would no longer be reabsorbed into the blood. Escaped plasma proteins return to the blood via the lymphatic system.

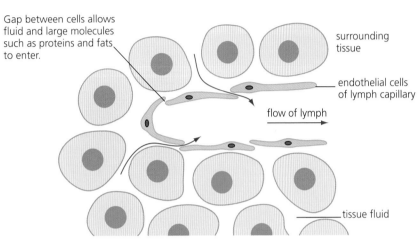

**Figure 9.7** Section through the end of a lymph capillary.

**Figure 9.8** A child with kwashiorkor, a condition caused by malnutrition.

The lymph capillaries drain the excess fluid through a system of larger vessels that connect with large veins beneath the collar bones where the fluid returns to the blood system. The lymph vessels have very thin walls and also have valves similar to those in veins. There are some muscle cells in the walls of the lymph vessels to move the fluid along.

Figure 9.8 shows a child suffering from severe malnutrition. The child's diet contains very little protein. Consequently his blood is very short of plasma proteins. One result is that fluid collects in the tissues, which is also called oedema. This causes the child's belly and limbs to swell, as you can see in the photo.

Figure 9.9 shows a woman suffering from a condition called elephantiasis. The woman has been infected with parasitic worms that have blocked the lymph vessels in one leg. Tissue fluid has accumulated in this leg, making it swell massively so that it looks like an elephant's leg.

**Figure 9.9** A person with elephantiasis, caused by parasitic worms that block lymph vessels.

> ### TEST YOURSELF
> **5** Explain why the escape of plasma proteins from capillaries could cause fluid to collect in the tissues.
> **6** Explain how a shortage of protein in the diet can lead to oedema.
> **7** Explain how a blockage in the lymph vessels, such as a parasitic worm, can cause oedema.
> **8** People with high blood pressure can have oedema in their hands and feet. Explain why.

159

# How blood is adapted to transport oxygen

Red blood cells have several adaptations for the transport of oxygen. Consider how each of the following features assists in efficient oxygen transport, before reading the explanations below. A red blood cell:

- is small in size: each cell is just over 7 μm in diameter. A blood capillary has an outer diameter of about 8 μm; the diameter of its lumen is rather less
- is shaped like a flattened disc
- has a thin central part of the disc
- has no organelles such as a nucleus or mitochondria
- is filled with haemoglobin.

## Small size

The small size allows the red cells to pass through the narrow capillaries. By being about the same diameter as the lumen of the capillary, they touch the sides, which reduces the distance that oxygen has to diffuse once it enters the capillary. The narrowness of the capillaries is an associated adaptation in that they can fit within the spaces between cells.

## Flattened disc shape

The flat shape (Figure 9.10) of a red blood cell increases the surface area to volume ratio and greatly increases the area through which oxygen can diffuse. It also results in all the haemoglobin being close to the surface, giving a short diffusion pathway. If the cells were spherical, much of the haemoglobin in the centre would be of little use because it would be too far for oxygen to reach it in the time available.

## Thin central part of disc

The thin centre allows the cell to be flexible so that it can bend and squeeze through the narrow capillaries. As blood flows through a capillary, the cell tends to form a dome shape with its edges scraping along the wall of the capillary.

## Absence of organelles

The absence of organelles provides maximum space for haemoglobin.

## Haemoglobin

The haemoglobin greatly increases the oxygen-carrying capacity of the blood. Oxygen is not very soluble in water, so only small amounts would be transported by plasma alone.

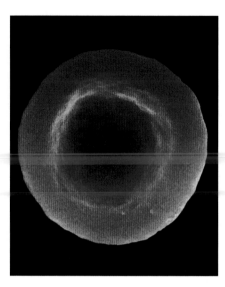

**Figure 9.10** The shape of a red blood cell (×15 000).

## TEST YOURSELF

9 Describe how you could separate the plasma and cells of a sample of blood.

10 Give two ways that red blood cells are adapted to oxygen transport.

11 Red blood cells do contain some enzymes. These are important for carbon dioxide transport. But red cells cannot replace the enzymes. Explain why the enzymes cannot be synthesised in the red cells.

# Haemoglobin

## The structure of haemoglobin

Haemoglobin is a **quarternary** protein (see Chapter 1, page 13) that consists of four polypeptides called globins, with a haem group tucked in the centre of each (Figure 9.11). It is the haem that is the key to the oxygen-carrying function of haemoglobin.

You will notice from Figure 9.12 that haem contains iron (symbol: Fe) at its centre. The iron exists as an $Fe^{2+}$ ion, which has the remarkable ability to combine reversibly with one oxygen molecule, making it ideal as a means of picking up and delivering oxygen, as we shall see later.

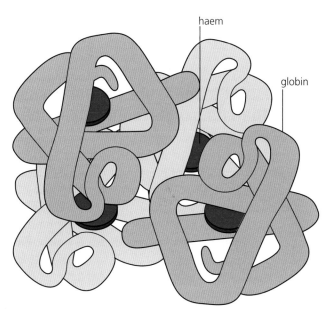

**Figure 9.11** Because haemoglobin contains four polypeptide chains, it is said to have a quarternary structure.

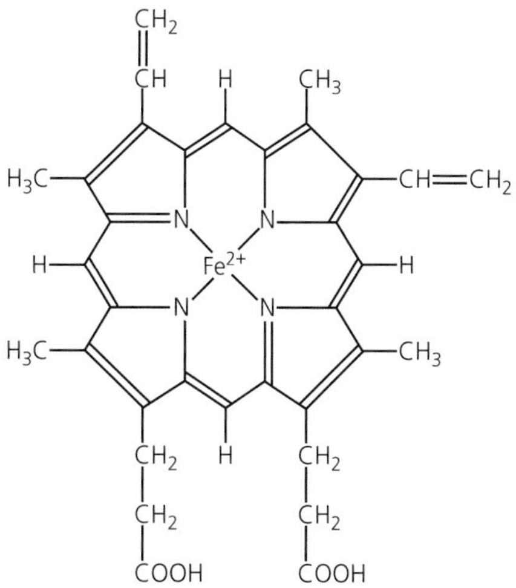

**Figure 9.12** The chemical structure of haem.

Haemoglobin molecules combine with oxygen when it is present in high concentrations, but – and this is the important feature – the process is reversed when the concentration is low. This means that adult human haemoglobin takes up oxygen in the lungs, but when it reaches a tissue where there is little oxygen the haemoglobin releases it again.

## Different kinds of haemoglobin

The structure of haem is the same in all haemoglobins, but the globin chains vary considerably between species. It is actually the globin component of haemoglobin that determines its precise properties, and many varieties exist.

Different forms of haemoglobin vary both in their oxygen-binding properties and in the conditions in which they take up and release oxygen. For example, llamas have haemoglobin suited to living at high altitude because it combines with oxygen more readily. This is useful because the partial pressure of oxygen at higher altitude is lower than normal. The haemoglobin of a developing baby in the womb differs from the haemoglobin the baby makes after its birth. Foetal haemoglobin is better at absorbing oxygen at low concentrations. This allows it to obtain its oxygen from the mother's blood, which has a much lower concentration than the air in the lungs.

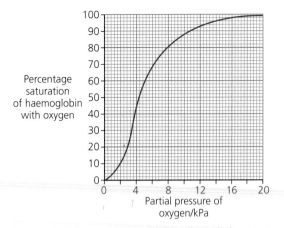

Percentage saturation of haemoglobin with oxygen

Partial pressure of oxygen/kPa

**Figure 9.13** The oxyhaemoglobin dissociation curve for adult humans.

After birth, the baby starts to make adult haemoglobin, which is better suited for the uptake of oxygen in the lungs.

Haemoglobins can associate and disassociate from oxygen, an exceptionally useful property.

## The oxyhaemoglobin dissociation curve

Figure 9.13 shows the oxyhaemoglobin dissociation curve for adult humans. This shows how much oxygen is combined with haemoglobin at different oxygen concentrations. However, if you look at the x-axis of the graph, you will see that the scale shows not a concentration but the 'partial pressure' of oxygen. Scientists use **partial pressure** as a measure rather than concentration because it is a more useful indicator of how much oxygen is available to haemoglobin.

## Extension

### Why is it more useful to measure partial pressure of oxygen in air than its oxygen concentration?

The air is a mixture of gases. Each gas exerts a pressure proportional to its concentration. Atmospheric pressure at sea level is about 100 kPa. (It varies a bit according to weather conditions, so forecasters refer to low-pressure areas and high-pressure areas.) The atmosphere is about 21% oxygen. So, at sea level, oxygen exerts a partial pressure of about 21% of the total atmospheric pressure; that is, about 21 kPa. Up a high mountain, the atmospheric pressure is much lower because the air is much thinner. But the proportion of oxygen is still about 21%. However, there is far less oxygen in each cubic metre of air, so the pressure exerted by the oxygen is much lower. On the summit of Mount Everest it is only about 7 kPa.

● What do we mean if we say that haemoglobin is 100% saturated?

In Figure 9.13, the y-axis shows the saturation of haemoglobin with oxygen. Each haemoglobin molecule can combine with a maximum of four molecules of oxygen, making oxyhaemoglobin. If all the haemoglobin molecules in a red blood cell combine with four oxygen molecules, the haemoglobin will be 100% saturated, and obviously it cannot possibly take on any more oxygen. (If 80% of the oxygen is released, the haemoglobin will then be only 20% saturated.)

**Oxyhaemoglobin** The complex formed when haemoglobin binds to oxygen.

### TIP

Oxyhaemoglobin dissociation curves often make more sense if you read them from right to left. The right-hand side of the curve is when the haemoglobin loads with oxygen at the lungs, then as the curve moves to the left it gradually unloads as the red blood cells reach parts of the body with a lower partial pressure of oxygen.

Let us look at the shape of the curve on the graph. It is like a partly flattened letter S. This is because the haemoglobin molecule changes shape and loads with oxygen more easily once the first oxygen has combined with it. For the first haem group to form a bond with the first oxygen, the four globins have to move a little relative to each other. Once they have moved, the next three oxygens can form bonds with the remaining three haem groups more easily. This is known as **cooperative binding** and explains why the first part of the S shape curves upwards. You will recall the similar idea of induced fit when enzymes bind with their substrate (see Chapter 3).

At the other end of the S shape, the curve flattens off. The graph shows that haemoglobin reaches nearly 100% oxygen saturation when the partial pressure is much lower than atmospheric partial pressure of 21 kPa.

● What is the advantage of this response to partial pressure?

In the alveoli of the lungs, the partial pressure of oxygen is less than the 21 kPa in the air outside the body. This is because the alveolar air contains a lot of water vapour and a relatively high concentration of carbon dioxide. The partial pressure of oxygen is usually about 15 kPa. From the graph you can see that even at this lower partial pressure the haemoglobin becomes almost 100% saturated.

Therefore, as the blood passes through the lung capillaries, the haemoglobin in the red cells becomes loaded with oxygen. This happens extremely quickly: the haemoglobin will become almost fully saturated within a fraction of a second. The blood system carries the red cells through a pulmonary vein into the heart from where they are pumped out through arteries and arterioles to the tissues of the body.

● Why does the oxyhaemoglobin keep its oxygen until it reaches the capillaries in the tissues?

The walls of the veins, arteries and arterioles are too thick to allow oxygen to escape rapidly. The partial pressure around the red cells remains constant, so the haemoglobin stays saturated. The data in Figure 9.14 explain why oxygen is unloaded from the oxyhaemoglobin when the red cells reach the capillaries in the tissues. When the red cells reach a capillary in, for example, a muscle or the brain, the surrounding tissue will have a low partial pressure of oxygen because these tissues will have been using oxygen for respiration.

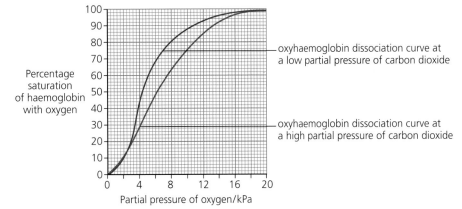

**Figure 9.14** The oxyhaemoglobin dissociation curve in low and high partial pressures of carbon dioxide.

Suppose that the partial pressure of oxygen in a muscle is 4 kPa. From the graph, at 4 kPa the haemoglobin can be no more than 55% saturated. Therefore, the oxyhaemoglobin rapidly unloads oxygen until the haemoglobin is 55% saturated. This unloading is called **dissociation**. Assuming that the haemoglobin became 100% saturated in the alveoli (in practice it is usually slightly less), it will unload 45% of its oxygen. This will rapidly diffuse into the muscle. Bear in mind that this all happens very fast as the blood circulates. (It is not like a truck that has to stop to load or unload.) So, although it may seem inefficient for oxyhaemoglobin to release only some of its oxygen, the muscle gets a continuous supply.

● Why is more oxygen unloaded in active muscle?

In practice, more than 45% of the oxygen is unloaded in active muscles. An active muscle is respiring and rapidly producing carbon dioxide. The data in the graph in Figure 9.14 explains this. Look at the curve for oxyhaemoglobin dissociation at a high partial pressure of carbon dioxide. You will see that it forms a more forward-sloping S. It is to the right of the curve for oxyhaemoglobin dissociation at a low carbon dioxide partial pressure. This change in position of the dissociation curve due to an increased concentration of carbon dioxide is called the Bohr effect. The increased concentration of carbon dioxide causes a change in the shape of the protein, in the same way that enzymes change shape with a change in pH.

You will recall that muscle has a 4% partial pressure of oxygen, when oxygen in oxyhaemoglobin above 55% saturation is unloaded into the muscle. Now read off the percentage saturation of haemoglobin when the partial pressure of carbon dioxide is high, as it will be in active muscle. It is only 30%. The oxyhaemoglobin will dissociate more completely and release 70% of its oxygen. The advantage of the Bohr effect is that when a muscle is respiring more it receives an increased oxygen supply.

---

EXAMPLE

## Oxyhaemoglobin dissociation in a foetus and its mother

As we have seen, there are different forms of haemoglobin in different animals, adapted for different environmental conditions or different stages in their lives. Figure 9.15 shows the oxyhaemoglobin dissociation curve for the type of haemoglobin that babies have before birth as well as for the mother's adult haemoglobin. While in the womb, a foetus receives its oxygen from the mother's blood by diffusion across the placenta.

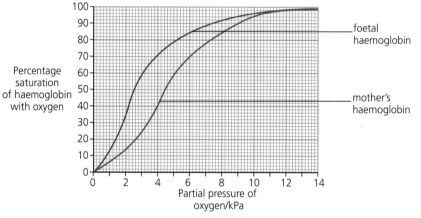

**Figure 9.15** Dissociation curves for foetal and adult oxyhaemoglobin.

1 The partial pressure of oxygen in the placenta may be 3 kPa. At that value, what percentage of the oxygen in the mother's oxyhaemoglobin will dissociate?
   *A line drawn up to the blue curve from 3 kPa will reach 24% on the percentage saturation axis.*
2 At the same partial pressure of 3 kPa, what percentage of the foetal haemoglobin will become saturated?
   *A line drawn up to the red curve from 3 kPa will reach 58% on the percentage saturation axis.*
3 Explain the advantage of the foetal haemoglobin having a different oxyhaemoglobin dissociation curve from the mother's haemoglobin.

*It means that the foetal haemoglobin can become saturated with oxygen even when the mother's haemoglobin is not fully saturated.*

A foetus starts to manufacture more and more adult haemoglobin as it gets closer to the time of birth. After birth, the proportion of foetal haemoglobin in the blood normally declines quite rapidly.

4 Explain why it is important for the baby to have adult haemoglobin after birth.
   *Once the baby is born, it is breathing in air with a comparatively high oxygen concentration and its haemoglobin will saturate with oxygen in the lungs. It needs adult haemoglobin so that the oxyhaemoglobin will gradually unload its oxygen at the different partial pressures of oxygen in the body.*

# The function of the heart

The heart is the most obviously active organ in the human body. It is constantly busy, even during sleep. When the heart stops beating we die. Traditionally the heart was thought to be the seat of our emotions: bravery, love, passion, excitement. This is not surprising since we can feel the changes in our heartbeat when we are frightened or get excited.

In reality, the function of the heart is relatively simple: to pump the blood round the body, forcing it to all parts of the body. Strictly, the human heart is two pumps, as it has two halves. One half pumps blood to the lungs, where it is oxygenated. The oxygenated blood returns to the other half of the heart and is then pumped to all the other organs.

The blood transports both oxygen and the glucose required for respiration. When the heart stops beating, the whole of the body is deprived of oxygen and fuel. Cells do not have reserves of oxygen. The brain cells are particularly sensitive to a shortage of oxygen. If the oxygen supply is stopped, within about 5 minutes brain cells will start to die. Rapid first aid to restart the heart is essential if brain damage is to be avoided.

The rate and strength of the heartbeat must be able to change according to the body's requirements. When we are at rest our hearts normally function at a steady 60–70 beats per minute. But, during exercise, when the muscles use oxygen at a much faster rate, the heart rate may more than double. In this chapter we will look at the structure and functions of the different parts of the heart.

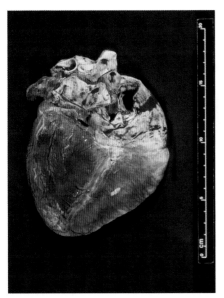

**Figure 9.16** The outside of the heart.

# The heart as a pump

Look at the diagram of the heart in Figure 9.17 (overleaf).

Notice that the heart has two halves, separated by a wall down the middle. Each half has two sections called chambers. The lower chambers are the ventricles. The right ventricle pumps blood to the lungs, and the left ventricle pumps it to the rest of the body. The two upper chambers are the atria (singular: atrium). The blood that returns from the lungs and the other parts of the body enters the atria. It is the job of the atria to collect and pump this blood into the ventricles.

**TIP**
The heart maintains a unidirectional flow of blood in the body. This helps to keep a steep concentration gradient at exchange surfaces, for example in the alveoli of the lungs (see Chapter 7).

**TIP**
Remember that diagrams show you the view from the front of a person, so you see the right side of the heart on the left of the diagram.

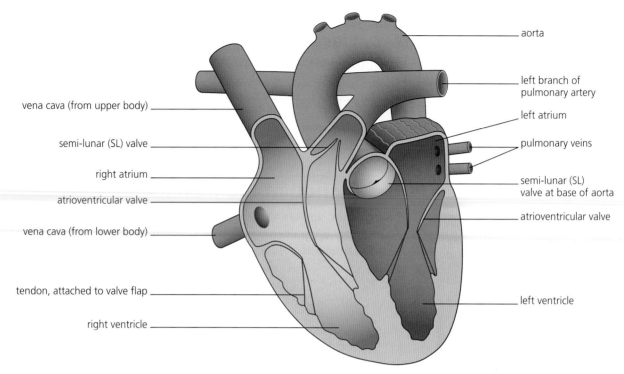

vena cava (from upper body)

semi-lunar (SL) valve

right atrium

atrioventricular valve

vena cava (from lower body)

tendon, attached to valve flap

right ventricle

aorta

left branch of pulmonary artery

left atrium

pulmonary veins

semi-lunar (SL) valve at base of aorta

atrioventricular valve

left ventricle

**Figure 9.17** A vertical section through the human heart.

**TIP**

Many students wrongly think that the wall of the left ventricle is thick to withstand the pressure in it. This is the wrong way round. It is thick to create the pressure needed to circulate the blood around the body.

**TIP**

Do not confuse the pulmonary artery and the pulmonary vein. The pulmonary artery, like all **a**rteries, carries blood **a**way from the heart. However, it is the only artery that carries deoxygenated blood. The pulmonary vein, like all ve**in**s, carries blood **in**to the heart. However, it is one of the few veins that carry oxygenated blood.

Remember: **a**rteries carry blood **a**way from the heart, and ve**in**s carry blood **in**to the heart.

Between the atrium and the ventricle on each side is a valve, called an atrioventricular valve (or AV valve). The valve that separates each atrium from its ventricle has flaps made of thin but tough tissue. The atrium muscles relax and the atrium fills with blood from the vena cava. Its muscles then contract. This pushes the flaps down, and blood flows into the emptied ventricle, which has relaxed muscles. Then, when the ventricle muscles contract, the flaps are pushed together. This stops blood from flowing back from the ventricle into the atrium.

In Figure 9.17 you can see that thin tendons join the edges of the valve flaps to the wall of each ventricle. These tendons are like tough pieces of string and do not stretch. The function of the tendons is to ensure that the valve only opens one way and the blood does not flow back into the atrium.

When the ventricles contract, blood is forced from the ventricles into large blood vessels that pass out of the top of the heart. From the left ventricle, blood enters the aorta. Outside the heart, the aorta has branches to the head, arms, intestines, legs, and so on. From the right ventricle, the pulmonary artery has a branch to each lung. Notice that there is a valve at the lower end of both the aorta and the pulmonary artery. These valves stop blood from flowing back into the ventricles when the ventricle muscles relax and the ventricles start to open up again to fill from the atrium.

Blood returns to the heart through large veins. The pulmonary veins return blood from the lungs to the left atrium. Veins called vena cavae bring blood from the upper and lower parts of the body into the right atrium.

Blood is pumped from the left ventricle to organs of the body where it becomes less well oxygenated. The blood returns to the right atrium. It then passes into the right ventricle, which pumps it to the lungs where it is re-oxygenated. It returns to the left atrium and then reaches the left ventricle again (see Figure 9.18).

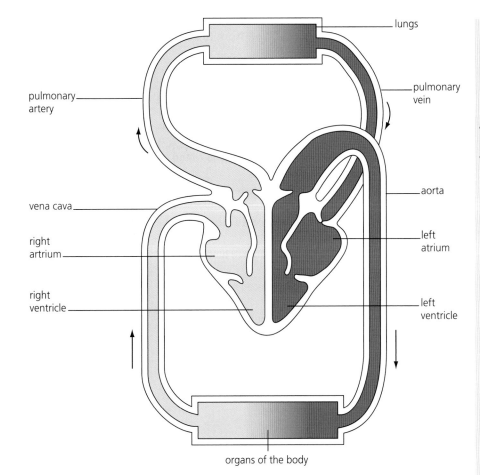

**Figure 9.18** The role of the heart in the mammal.

## REQUIRED PRACTICAL 5

### Dissection of animal or plant gas exchange system or mass transport system or of organ within such a system
This is just one example of how you might tackle this required practical.

### Heart dissection
Figure 9.19 shows the heart as seen from the front (an anterior view)

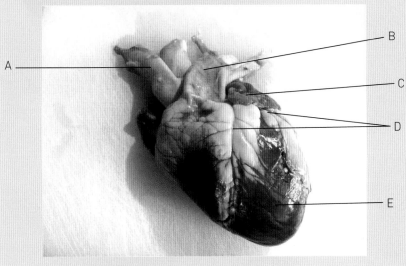

**Figure 9.19** The anterior view of the heart.

Figure 9.20 shows the heart as seen from the back (a posterior view)

**1** Identify structures A–K as shown on Figures 9.19 and 9.20.
**2** Copy and complete the table.

| Structure | Function |
|---|---|
| | Receives deoxygenated blood from the body |
| | Carries oxygenated blood round the body |
| | Pumps deoxygenated blood to the lungs |
| | Takes deoxygenated blood from the body to the heart |
| | Receives oxygenated blood from the lungs |
| | Pumps oxygenated blood from the heart |
| | Carries oxygenated blood from the lungs back to the heart |
| | Carries deoxygenated blood from the heart to the lungs |
| | Carry oxygenated blood to the heart muscle |

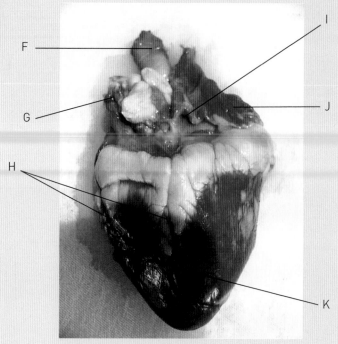

**Figure 9.20** The posterior view of the heart.

Figure 9.21 shows the atrioventricular valve (AV) as seen in the right ventricle. Notice the flaps of the valve and the tendons attached to them.

**3** Explain how these strong tendons help the valve to work effectively.

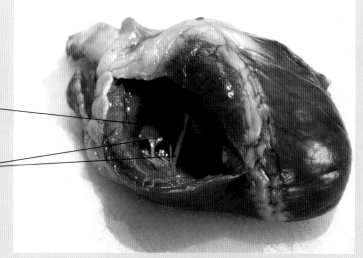

Flap of AV valve

Tendons attached to valve (chordae tendinae)

**Figure 9.21** The AV valve in the right ventricle.

Figure 9.22 shows the left ventricle and the left AV valve. Notice that the wall of the left ventricle is much thicker than the wall of the right ventricle.

**4** The wall of the left ventricle is much thicker than the wall of the right ventricle. Explain the advantage of this.

**Figure 9.22** The left ventricle and the left AV valve.

Figure 9.23 shows a red pencil inserted into a blood vessel and passing through to the left ventricle.

**5** Name the blood vessel and the structures the pencil passes through.

In Figure 9.24 the green pencil has passed from the left ventricle into a blood vessel.

**6** Name the structures it passes through.

**Figure 9.23** A pencil shown inserted into a blood vessel and passing through to the left ventricle.

**Figure 9.24** A pencil passing from the left ventricle into a blood vessel.

In Figure 9.25 the lilac pencil is inserted into a blood vessel that enters a chamber of the heart.

**7** Name the blood vessel and the chamber of the heart it enters.

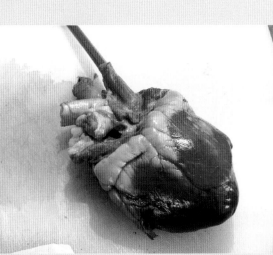

**Figure 9.25** A pencil is inserted into a blood vessel that enters a chamber of the heart.

In Figure 9.26 the yellow pencil is passing from one chamber of the heart to another, via a valve.

**8** Name the chambers and the valve.

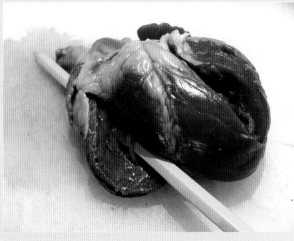

**Figure 9.26** A pencil passes from one chamber of the heart to another, via a valve.

# The cardiac cycle

Since you are reading this, you must be alive and your heart must be beating. To confirm this, gently press a fingertip on your neck just to one side of your trachea. Here, each time the heart beats, you should feel the pulse caused by the surge of blood through one of the arteries that goes to your head. If you are relaxed, you will feel a pulse roughly every second. The time between each pulse represents the length of the cardiac cycle.

> **Cardiac cycle** The sequence of events that make up one heartbeat.

The cardiac cycle is the sequence of stages that happens during one heartbeat. The term for a stage in the cardiac cycle when the muscles of the heart chambers are contracting is **systole**. The atria contract during atrial systole and the ventricles contract during ventricular systole. The relaxation stage is called **diastole**. Therefore, the three stages shown in Figure 9.27 are atrial systole, ventricular systole and diastole.

**TIP**

You do not need to be able to recall the terms 'systole' and 'diastole', but if you do use them, make sure you get them the right way round.

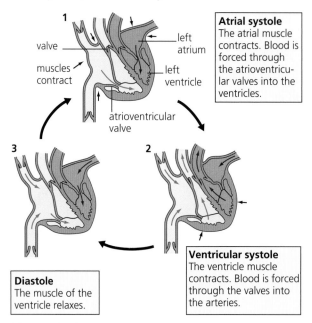

**Atrial systole**
The atrial muscle contracts. Blood is forced through the atrioventricular valves into the ventricles.

**Ventricular systole**
The ventricle muscle contracts. Blood is forced through the valves into the arteries.

**Diastole**
The muscle of the ventricle relaxes.

**Figure 9.27** The three stages in the cardiac cycle. The atria contract (atrial systole), the ventricles contract (ventricular systole) and finally the ventricles relax (diastole).

Contraction of the muscles in the walls of the ventricles creates the pressure that circulates the blood. The left and right ventricles contract at the same time.

As the ventricles contract they build up a pressure higher than that in the aorta or the pulmonary artery, which forces open the valves at the base of the aorta and the pulmonary artery and drives blood out through these vessels. When the ventricle muscle relaxes, the ventricle walls recoil. This increases the volume of the ventricle and reduces the pressure so it is lower than the pressure in the aorta and the semi-lunar valve shuts. This means blood doesn't flow back into the ventricle. The pressure in the ventricle then falls below that in the atrium, the atrioventricular valve opens and blood flows into the ventricle.

After passing round the lungs or the rest of the body, the blood flows back into the atria of the heart at a much lower pressure. When the blood flows back in, the muscular walls of the heart are relaxed, so even at low pressure the returning blood expands the relatively thin walls of the atria. As the atria fill, some blood does pass through the valves into the ventricles. However, as the atria fill up, the muscles in their walls contract. This forces the valves to open fully and pushes the remaining blood quickly into the ventricles.

Each time the heart beats, the left ventricle pushes out the same volume of blood as the right ventricle. This may seem surprising, since the lungs are much nearer and smaller than the rest of the body, but otherwise the continuous circulation would not be maintained. If the right ventricle pumped out less blood each time, there would be less returning to the left atrium and therefore less for the left ventricle to pump to the body. The left ventricle would soon run out of blood to send to the body.

## Analysing the pressure changes and valve openings and closings in the heart during the cardiac cycle

The graph in Figure 9.28 shows the changes in pressure that occur in the left side of the heart and in the aorta during one cardiac cycle. Let us look first at the curve showing pressure in the left ventricle. Key points on the curve are labelled with the letters A to D.

**Figure 9.28** A graph showing changes in pressure during one cardiac cycle for the left atrium and left ventricle and the aorta.

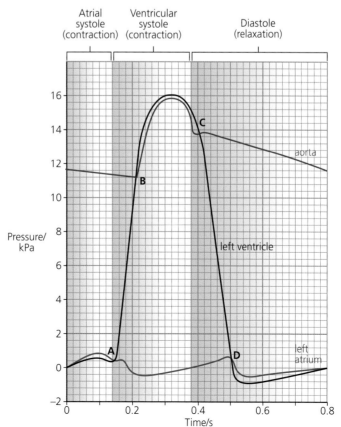

171

- Notice that at the start of atrial contraction the pressure in the left ventricle is more or less 0 kPa. At this time the muscles in the walls of the ventricle are relaxed and therefore not putting any pressure on the blood. During atrial contraction the pressure rises slightly as the left atrium contracts and pushes blood into the ventricle.
- At point A, the start of ventricular contraction, the ventricle walls contract strongly and the pressure shoots up. The pressure increase forces the atrioventricular valve to shut, as the pressure in the ventricle is greater than the pressure in the atrium, preventing blood from being pushed back into the atrium.
- At B, the pressure in the ventricle becomes the same as the pressure in the aorta. As soon as the pressure in the ventricle exceeds the pressure in the aorta, the valve at the base of the aorta is forced open and blood is pushed into the aorta.
- By C, the ventricle has been emptied of blood. The muscles in the ventricle wall relax, pressure in the ventricle falls back to below zero, i.e. lower than the pressure in the aorta, and the aortic valve shuts due to the greater pressure in the aorta.
- At D, for a short time there is a slight period of lower pressure as the ventricle expands and its internal volume increases.

Remember that one cardiac cycle follows another in a continuous process. There is never a gap between cycles, so blood immediately starts to flow into the expanding ventricles from the atria because the pressure in the atria is now greater than the pressure in the empty, expanding ventricles. This in turn causes blood to flow into the atria from the veins. The pressure in the ventricles rises again as they fill.

## EXAMPLE

### Pressure changes during the cardiac cycle

Look at the curves for the left atrial and aortic pressures shown in Figure 9.28.

1 Describe and explain what happens to the pressure in the atrium between the start of the cycle and A.
*The pressure rises as the muscles in the atrium wall contract and force blood through the valve into ventricle. The pressure falls again as the atrium empties.*

2 The atrioventricular valve closes at A. When does it open again? Explain your answer.
*The valve opens at D. Notice that the curves for pressure in the ventricle and atrium cross at this point, so the pressure on either side of the valve is briefly the same. The pressure in the ventricle then falls slightly below the atrial pressure, so blood again flows through the valve into the ventricle.*

3 Soon after A, the pressure falls as the atrium expands after atrial systole. Explain why the pressure then rises again until D.
*The pressure of blood in the pulmonary vein, even though it is quite low, is pushing blood into the atrium and raising the pressure in the atrium as it fills up.*

4 Explain why the maximum pressure in the atrium is much lower than the maximum pressure in the ventricle.
*There is much less muscle in the walls of the atria. They exert much less force when they contract compared with the thick walls of the ventricle.*

The pressure in the aorta stays high throughout the cycle. Its walls are elastic and are stretched during ventricular systole. The stretched walls exert pressure on the blood in the aorta, just as an elastic bandage on your leg continues to squeeze the leg.

5 Between which points on the curve is blood entering the aorta from the left ventricle?
*Blood enters the aorta between B and C. The valve opens at B and closes again when the pressure in the ventricle falls below the pressure in the aorta.*

6 Figure 9.28 shows the pressures on the left side of the heart. How would you expect a similar graph showing pressures on the right side to differ? Explain your answer.

It may help you to refer to the drawing of the heart in Figure 9.17.

*The maximum pressure in the right ventricle would be lower because the walls are less muscular and exert less force. In fact, the maximum pressure in the right ventricle is normally less than a quarter of the maximum in the left ventricle. The pressure in the pulmonary artery would also be much lower. However, since the two sides pump together, the timing and pattern of events would be much the same.*

## Calculating cardiac output

Figure 9.29 shows the internal volume of the left ventricle during one cardiac cycle. This graph is on exactly the same time scale as the graph of pressure changes in Figure 9.28. The same positions have been labelled A–D, so you can match up with the changes in pressure in the left ventricle.

You can see that between B and C the volume of the ventricle decreases rapidly. This matches the period when pressure increases as the ventricle contracts. During this stage, blood is being pumped out into the aorta. As the muscles of the ventricle relax after C, its volume increases, the pressure falls and blood flows in from the atrium. At this stage, the muscles of the atrium are still relaxed. Atrial systole, when atrial muscles contract, merely tops up the blood in the ventricle between zero and point A on the curve in this graph.

1 Notice in Figure 9.29 that the volume of the left ventricle stays almost constant between A and B. How can this be explained?
*This is explained because the atrioventricular valve and the aortic valve are both shut. Therefore no blood is entering or leaving the ventricle so its volume stays the same.*

The volume of blood pumped out of the left ventricle during one cardiac cycle is called the stroke volume.

2 Use the scale on the *y*-axis in Figure 9.29 to calculate the stroke volume shown by this graph.
*This is about 75 cm³.*

We call the volume of blood that the left ventricle pumps out to the body per minute the cardiac output. To work out the cardiac output, you first need to calculate the heart rate, which is the number of cardiac cycles per minute. To find the number of cycles per minute you have to divide 60 by the time in seconds taken for one cycle.

3 How long is the cardiac cycle shown in Figure 9.29?
*One cardiac cycle lasts 0.8 seconds. You can then use the following equation to find the cardiac output:*

Cardiac output = stroke volume × heart rate

4 Calculate the cardiac output shown in Figure 9.29.
$$\text{Heart rate} = \frac{60}{0.8} = 75$$
Cardiac output = 75 × 75 = 5625 cm³

*For a man lying down and doing nothing, the average cardiac output is about 5 dm³ (5000 cm³). During exercise the cardiac output can increase to be about four times as great. This caters for the much higher oxygen and glucose requirements of active muscles. The increase can be produced by changing either the heart rate or the stroke volume, or both.*

In an investigation, groups of trained athletes and untrained students were required to ride an exercise bike as fast as they could for a few minutes. Their maximum stroke volume and heart rate were measured. The mean maximum stroke volume for the athletes was

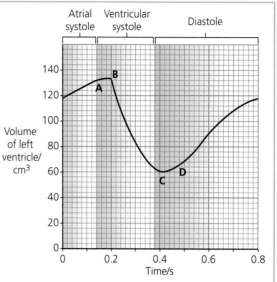

**Figure 9.29** Graph of volume changes in the left ventricle in one cardiac cycle.

160 cm³ and for the students 100 cm³. The mean maximum heart rate was 190 beats per minute for the athletes and 200 beats per minute for the students.

5 a) Calculate the mean maximum cardiac output for each group.
 b) Explain which factor accounts for the difference.
 c) Suggest how training may have affected the stroke volume.
*a) For the athletes the mean maximum cardiac output = 160 × 190 = 30 400 cm³. For the students mean maximum cardiac output = 100 × 200 = 20 000 cm³.*
*b) The stroke volume accounts for the difference. Both groups have a similar maximum heart rate, but the athletes have a greater stroke volume so they can circulate more oxygenated blood round the body in a minute.*
*c) Training has increased the strength of the muscle in the wall of the left ventricle. During exercise, more deoxygenated blood returns to the right side of the heart to be oxygenated at the lungs. In turn, this leads to more oxygenated blood entering the left side of the heart. Therefore the ventricles are fully filled and this stretches their walls. As a result, the ventricles contract more forcefully.*

### TIPS
- You do not need to learn the formula for cardiac output or memorise the terms 'cardiac output' or 'stroke volume'.
- To find how long one cardiac cycle is on a graph with more than one cycle, find an easily identifiable point on the curve and measure the time taken to reach the same point on the next cycle.
- 1 dm³ is 1000 cm³.

# Extension

## Coronary heart disease

**Coronary heart disease** (or CHD) refers to any condition that interferes with the coronary arteries that supply blood to the heart muscle. Being very active, the muscle requires a continuous supply of oxygen and glucose. This supply does not enter the muscle from the blood in the chambers of the heart. It reaches the muscle from arteries that branch from the aorta and spread over the surface of the heart. Small vessels penetrate the heart muscle so that all parts are close enough to the blood supply for the oxygen and glucose to diffuse to the muscle cells.

Problems arise when blood vessels taking blood to the heart muscle become very narrow or blocked so that the supply of oxygen and glucose is reduced or cut off. How serious this is depends on where the blockage occurs and how much of the muscle is affected.

Arteries may become partly blocked by the formation of fatty deposits in their walls when the layer of cells lining the inside of the artery becomes damaged and inflamed. These fatty deposits are called atheroma. The resulting narrowing of the arteries is referred to as atherosclerosis or, in common language, 'hardening of the arteries'. Blood clots often form in a narrowed artery, a condition that is called **thrombosis**. If the blood supply to a major part of the heart muscle is completely blocked, the muscle starts to die due to lack of a blood supply. A heart attack follows. The medical term for heart muscle death is myocardial infarction (an infarction is the death of tissue; myocardial means that it occurs in the heart muscle).

## EXAMPLE

### Trialling aspirin as a treatment to prevent heart attacks

It was first suggested in the 1960s that aspirin might help to prevent heart attacks when it was discovered that aspirin helps to stop blood clots. This is because aspirin interferes with the action of blood platelets. The platelets are cell fragments that play an important role in blood clotting. Aspirin stops the platelets from sticking together and so reduces the chance that clots will block the vessels taking blood to the heart muscle.

At the time, most doctors were very scornful of the idea that such a cheap and common drug as aspirin could prevent heart attacks. Nevertheless, the Medical Research Council tested the idea, and in 1969 it set up a trial in Cardiff. The researchers decided to test the effect of giving a daily dose of aspirin to men who had recently recovered from a heart attack. One group of the men took a daily pill containing 300 mg of aspirin. The others, the control group, took a dummy pill containing no aspirin. The patients in each group were selected randomly, so neither the doctors nor the patients knew who was being treated with aspirin. Over 1200 men were studied, but only the coded records of the researchers enabled them

to trace which of the men had taken the aspirin. After treatment for a year, 12% fewer of the men taking the aspirin had died of another heart attack compared with the control group. However, the difference between deaths in the two groups was not enough for the results to be statistically significant. This one study had not given convincing evidence for doctors to start using aspirin as a regular part of the treatment for heart attack victims.

The research continued. Evidence was gathered from larger numbers, for longer times after the heart attack, from women as well as men, for different age groups and with different doses of aspirin. The results from these studies were collected and analysed. All showed that treatment with aspirin gave some benefit. In fact, the combined results of about 20 years of studies involved large enough numbers to show that aspirin gave a significant advantage after a first heart attack. During these studies, compared with the control groups, on average 34% fewer of the patients taking a daily dose of aspirin died.

It is now common for people who have had a heart attack to take aspirin. The evidence shows that this simple treatment prevents, or at least delays, many thousands of deaths a year. At the same time, this example illustrates some of the difficulties of developing new treatments in medicine. People vary enormously and respond differently to drugs and other treatments. Trial results rarely show absolutely clear-cut benefits, and some people may have unexpected side effects.

1 The men gave permission to take part in the first trial, but they could not choose to receive the treatment that might help to save their lives.

  a) Do you think they should have been allowed to choose?

  *The men had to give permission to receive the aspirin treatment, as they needed to be aware that there might be side effects. It might seem fairer to let the men choose, but if they had been able to choose, this could have damaged the validity of the study.*

  b) How might the results of the trial have been affected if they had been able to choose?

  *The men needed to be put into groups at random. The kind of person who wants to try the aspirin treatment might be different from those who prefer to have the placebo. For example, they might be more interested in looking after their health. This would mean the two groups would not be random. Also, if a person knows they are receiving a drug treatment, they may actively look for benefits and report these to the doctors carrying out the study.*

2 How should dummy pills be made in order to make sure the results are reliable?

*They should contain all the ingredients of the aspirin tablets except the aspirin, so that there is only one variable. They should also look the same as the aspirin tablets so that the people taking them cannot tell whether they are the real tablets or placebo.*

3 Why do you think the doctors were not told which patients were getting the aspirin?

*Even though doctors try to be impartial, it is very difficult for them not to look for benefits of a drug if they think it helps patients. They may be more proactive in noticing health benefits in people taking aspirin, if they know which patients these are.*

Aspirin can have side effects. For example, if a person also develops a stomach ulcer, it may bleed excessively because the aspirin reduces clotting. Yet the improvements in the survival rate after a heart attack show that the benefits from aspirin treatment are much greater than the risks from this side effect. However, because the benefits of aspirin were publicised, some people started taking aspirin to prevent a heart attack, even though they had never either suffered from one or shown signs of having one.

4 Suggest what advice could be given to the public and to doctors about this use of aspirin.

*This is a 'suggest' question so you need to give your own point of view, backed up with reasons. For example, you may think that people should be informed about the side effects of taking aspirin, so that they understand that it is not beneficial to take aspirin every day unless you have been recommended to do so by your doctor.*

**TIP**

You don't need to learn the risk factors for coronary heart disease. Make sure you can interpret data concerning these risk factors and analyse it though.

# Extension

## Risk factors for CHD

Many factors increase the risk of suffering from coronary heart disease. Some of these factors are unavoidable, such as increasing age. But others are the result of particular lifestyles, and a person can choose to alter the risk from them. The main risk factors include the following.

- **Age and sex** Deaths from CHD can occur in young adults, but the numbers involved are small. Not surprisingly, the risk increases with age as damage to arteries develops slowly and as the effects of other factors take effect. Men are much more likely than women to get CHD in middle age, but after this the risk becomes fairly similar.
- **Genetic factors** CHD tends to run in families, especially where heart attacks occur in middle age or earlier. This may be partly due to members of families having similar lifestyles, but there is evidence for some genetic causes. For example, the risk of two identical twins both dying from CHD is about four times as great as the risk for non-identical twins.

- **Smoking** As we saw in Chapter 7, in addition to its effects on the lungs, smoking can significantly increase the risk of dying from CHD. Table 7.1 on page 128 shows that the risk for doctors who smoked heavily was nearly double that of non-smokers. Working out the precise effect is difficult. For heavy smokers it is likely that smoking will damage both the lungs and the heart. It may be chance as to which is the first to cause death. It is not certain how smoking affects the heart. One possible explanation is that nicotine makes arteries constrict, and this causes an increase in blood pressure, another risk factor.

- **High blood pressure** It is normal for blood pressure to increase during exercise, when the heart beats more forcefully. However, in some people the pressure is high even when they are at rest. The risk factors linked to high blood pressure include genetics, high salt intake, lack of exercise and alcoholism. One effect of high blood pressure is that the arteries develop thicker walls. As the wall of an artery thickens, the lumen gets narrower. As anyone who has attached rubber tubing to a tap and then squeezed will know, the water jet comes out with much greater force. The narrowing of the arteries therefore has the knock-on effect of raising blood pressure even more. This can damage their inner surface, making it more likely that atheroma will develop. It can also result in damage to the heart itself. The ventricles can enlarge, and in the worst cases the beating of the heart can become so irregular that heart failure results.

- **High concentration of low-density lipoproteins in the blood** Low-density lipoproteins (LDL) are involved in the formation of atheroma. Lipoproteins are a complex association of triglycerides, cholesterol and proteins. Although cholesterol has a bad reputation because of its link to heart disease, it is essential for the synthesis of cell membranes. Cholesterol is transported in the blood by LDL from the liver where the lipoproteins are made from fats and cholesterol in the diet. However, research shows that when the diet contains an excess of fats, especially saturated fats, the quantity of LDL in the blood rises and there is a greatly increased risk of developing coronary heart disease. The blood also contains high-density lipoprotein (HDL), which has a higher proportion of protein in their structure. These are the 'goodies' because they absorb excess cholesterol and return it to the liver where it is removed from the blood. People with a high ratio of HDL to LDL have a lower risk of developing heart disease. Blood tests can determine the ratio of HDL to LDL, and the results can be used to advise on protective changes to lifestyle. Drugs called statins may also be used to lower the concentration of LDL.

## Measuring blood pressure

Although scientists use kilopascals (kPa) as the units of pressure, doctors still measure blood pressure in old units. These were the units in which a mercury barometer measured atmospheric pressure. Mean atmospheric pressure is 760 millimetres of mercury (mmHg), which is the height of a column of mercury that atmospheric pressure will support. Atmospheric pressure is 100 kPa, so 760 mmHg = 100 kPa.

Blood pressure is measured by finding the maximum and minimum pressures in the artery in the arm. In an adult, a healthy reading is taken to be less than 140 and 90 mmHg for these two measurements.

## Risk factors work together

There are some other factors that show a statistical increase in the risk of heart disease, such as obesity, lack of exercise and diabetes. In some cases, the explanation for the increased risk is clearly linked to other factors. For example, obesity is likely to be linked to eating fatty foods, which in turn is likely to lead to high concentrations of cholesterol and LDL in the blood. For many people, their statistical risk of having a heart attack is the result of a combination of different risk factors.

## ACTIVITY

### Calculating risk of CHD

For each risk factor, statisticians can assess the chance of developing coronary heart disease. Table 9.1 gives points for some risk factors for men in different age groups. The points are based on statistical evidence for the increased risk for healthy men over a 10 year period. For example, the point score of a man aged between 40 and 49 is increased by 5 points if he is a smoker. His total point score in Table 9.1 shows the probability that he will develop CHD during the next 10 years of his life, assuming that he is healthy at the time of assessment.

To calculate the point score, you add together the scores for all of the five categories in Table 9.1. For example, a 42-year-old non-smoking man, with a cholesterol concentration of $180\,mg\,100\,cm^{-3}$ of blood, blood pressure of $18.7\,kPa$ and an HDL concentration of $55\,mg\,100\,cm^{-3}$, has these scores:

| | |
|---|---|
| Age | 3 |
| Cholesterol | 3 |
| Smoking | 0 |
| Blood pressure | 1 |
| HDL | 0. |
| The total score is | 7. |

From Table 9.2, you can see that this man has a 3% increased risk of suffering from CHD during the following 10 years. This is called his '10 year risk'.

1 Suggest explanations for each of the following.
  a) The point score rises steeply for older age groups.
  b) The point score for blood cholesterol is greatest for the youngest age group.
  c) Blood HDL content above $60\,mg\,100\,cm^{-3}$ has a negative score.

2 Work out the 10 year risk for each of the following people and suggest what advice about their lifestyle should be given in each case.
  a) A male smoker, aged 48, blood cholesterol $265\,mg\,100\,cm^{-3}$, blood pressure $19.5\,kPa$ and HDL $44\,mg\,100\,cm^{-3}$.
  b) A male non-smoker, aged 68, blood cholesterol $188\,mg\,100\,cm^{-3}$, blood pressure $18.7\,kPa$ and HDL $62\,mg\,100\,cm^{-3}$.

**Table 9.1** Estimates of the risk of coronary heart disease over the next 10 year period for different age groups of men.

| Factor | Range | Points for each age group Age groups/years | | | |
|---|---|---|---|---|---|
| | | 40 to 49 | 50 to 59 | 60 to 69 | 70 to 79 |
| Age | | 3 | 7 | 10 | 12 |
| Blood cholesterol/mg $100\,cm^{-3}$ | <160 | 0 | 0 | 0 | 0 |
| | 160–199 | 3 | 2 | 1 | 0 |
| | 200–239 | 5 | 3 | 1 | 0 |
| | 240–279 | 6 | 4 | 2 | 1 |
| | 280+ | 8 | 5 | 3 | 1 |
| Smoking | Smoker | 5 | 3 | 1 | 1 |
| | Non-smoker | 0 | 0 | 0 | 0 |
| Systolic blood pressure/kPa | <16 | 0 | 0 | 0 | 0 |
| | 16–17.2 | 0 | 0 | 0 | 0 |
| | 17.3–18.5 | 1 | 1 | 1 | 1 |
| | 18.7–21.2 | 1 | 1 | 1 | 1 |
| | 21.3+ | 2 | 2 | 2 | 2 |
| Blood HDL content/mg $100\,cm^{-3}$ | <40 | 2 | 2 | 2 | 2 |
| | 40–49 | 1 | 1 | 1 | 1 |
| | 50–59 | 0 | 0 | 0 | 0 |
| | 60+ | −1 | −1 | −1 | −1 |

**Table 9.2** Point scores for men and the percentage probability that they will develop CHD in the next 10 years.

| Point score | 1 | 2 | 3 | 4 | 5 | 6 | 7 | 8 | 9 | 10 | 11 | 12 | 13 | 14 | 15 | 16 | 17+ |
|---|---|---|---|---|---|---|---|---|---|---|---|---|---|---|---|---|---|
| Percentage increased risk of CHD over next 10 years | 1 | 1 | 1 | 1 | 2 | 2 | 3 | 4 | 5 | 6 | 8 | 10 | 12 | 16 | 20 | 25 | 30 |

# Practice questions

**1** The lugworm is an animal that lives in sandy beaches. It lives in a tube, obtaining oxygen from sea water, which passes through the tube. However, when the tide goes out, very little oxygen is available. Here is the dissociation curve for lugworm oxyhaemoglobin.

**a)** Explain the advantage to the lugworm of having an oxyhaemoglobin dissociation curve of this kind. *(2)*

**b)** Haemoglobin has a quaternary structure. What does this mean? *(1)*

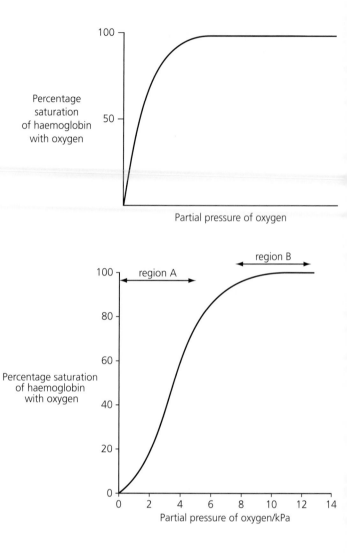

**2** The graph on the right shows the dissociation curve for human oxyhaemoglobin at a low carbon dioxide concentration.

**a)** Explain the advantage of the part of the oxyhaemoglobin dissociation curve in:

**i)**   region A *(1)*

**ii)**  region B. *(1)*

**b) i)**  Sketch a line on the figure to show the shape of the oxyhaemoglobin dissociation curve that you would expect if the concentration of carbon dioxide was higher. *(2)*

**ii)**  Explain the advantage of the effect of carbon dioxide on the shape of the oxyhaemoglobin dissociation curve. *(2)*

**3** The graph shows the change in volume of the left ventricle during one complete cardiac cycle in one individual.

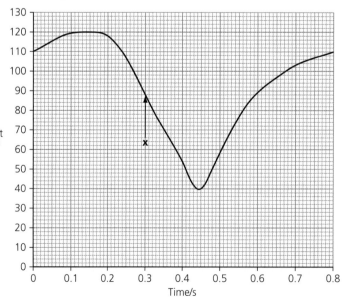

**a)** Copy and complete the table to show whether the following valves are open or closed at time X on the graph. *(1)*

| Valve | Open or closed? |
|---|---|
| Atrioventricular valve | |
| Semi-lunar valve | |

**b)** **i)** The stroke volume is the volume of blood pumped out of the left ventricle in one cardiac cycle. Use the graph to calculate the stroke volume of this person's heart. *(2)*

**ii)** The cardiac output is the volume of blood pumped out by the left ventricle in 1 minute. Calculate this person's cardiac output. Show your working. *(2)*

**4** Scientists carried out a study to investigate the effects of giving up smoking on the incidence of heart disease. They used large numbers of people who were divided into three groups:

- people who had never smoked
- people who had given up smoking
- people who continued to smoke regularly.

All the people were free of heart disease at the start of the study. They checked the health of each person at regular intervals and tested them for heart disease. The results are shown in the graph.

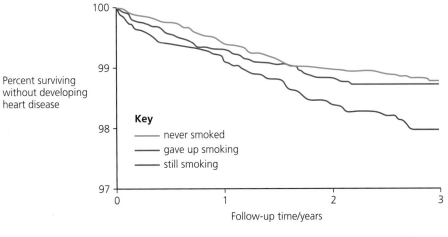

**a)** Describe the results of this investigation. *(3)*

**b)** The scientists who carried out the study said 'It's never too late to give up smoking. The risk of heart disease reduces by 50% after 1 year of giving up smoking.' Evaluate this statement. *(4)*

**5** The diagram shows some of the risk factors associated with deaths from coronary heart disease (CHD) in people under the age of 75 in the UK during 1998.

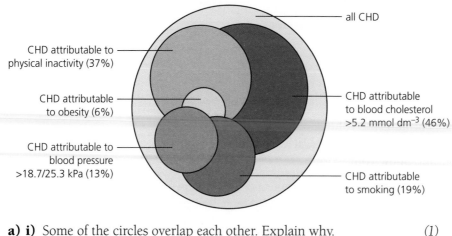

**a) i)** Some of the circles overlap each other. Explain why. *(1)*

**ii)** Some deaths from CHD are not related to any of the risk factors shown in the diagram. Suggest one possible cause of these deaths. *(1)*

**b)** Beta-blockers are drugs that reduce blood pressure. Doctors carried out an investigation to find the effect of different doses of a beta-blocker on blood pressure. They used a large number of people who all had high blood pressure in their investigation. The graph shows the results.

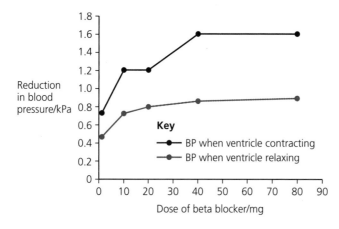

**c)** What dose of the beta-blocker should be used to treat people with high blood pressure? Give reasons for your answer. *(3)*

## Stretch and challenge

**6** You have seen the pattern of the circulation in a mammal in this chapter. Evaluate the effectiveness of this sort of circulation compared to the pattern of circulation found in fish.

# 10

# Mass transport in plants

**PRIOR KNOWLEDGE**

*Before you start, make sure that you are confident in your knowledge and understanding of the following points.*
- Plant cells have a cell wall made of cellulose, which strengthens the cell.
- When there is a water potential gradient, water moves across cell membranes by osmosis.
- A tissue is a group of cells with similar structure and function. Examples of plant tissues include xylem and phloem, which transport substances around the plant.
- Plant growth hormones are used in agriculture and horticulture as weedkillers.
- Environmental factors affect the rate of transpiration.

**TEST YOURSELF ON PRIOR KNOWLEDGE**
1 What does xylem transport in plants?
2 What is the advantage of using a plant growth hormone as a weedkiller?

## How does water get to the top of a tree?

An oak tree can be at least 30 m high and have a huge array of leaves and branches. Tall trees like this transport water and other materials to the crown from the soil over a vast distance. You may have seen a physics experiment using a pump in which water is 'sucked' up a long tube, perhaps up the side of a school building. No matter how good the pump, the water never goes up more than about 10 m. So, how can a tree get water to the leaves at the top?

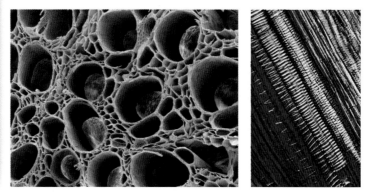

**Figure 10.1** (a) Cross-section of a stem showing the xylem tissue. (b) Vertical section of xylem.

In the stem of a plant is the specialised transport system through which water moves rapidly upwards to the leaves. The **xylem vessels** are stained red (Figure 10.1a).

## Extension

Notice that the xylem vessels in Figure 10.1a appear to be empty. This is because the vessels are long tubes consisting only of cell walls linked end to end and containing no cytoplasm. They are created when living cells die, leaving only the cell walls. Near the tip of a young root they develop as elongated cells. The cellulose walls of these cells become thickened with a waterproof substance called **lignin**. At first, the lignin is in rings, as you can see in Figure 10.1b. The rings allow the cells to stretch and grow longer. The spaces between rings also allow water to enter the cells easily. In older cells, the lignin fills in the spaces between the rings so that the walls are almost completely lignified, apart from small gaps, called **pits**, that allow water to move

sideways into surrounding tissues or between vessels if any get blocked. The end walls of the original cells break down, so the vessels are continuous pipes going from the roots all the way up the trunk of a tree to the uppermost leaves.

Some of the tallest trees in the world are conifers, which have tracheids instead of xylem vessels. Tracheids are much smaller, do not have perforated ends, and so are not joined end-to-end into other tracheids like xylem vessels. The pits in tracheids are the only way water can move from one tracheid to another, maintaining a continuous column of water up the tree.

## Pulling water up

Transpiration The evaporation of water vapour, mainly through the stomata in the surface of a leaf.

To understand how water reaches the top of the tree, we need to go to the leaves. In Chapter 7 we looked at the structure of leaves. You will remember that leaves have stomata, which open to let in carbon dioxide for photosynthesis. An unavoidable result of this is that water can diffuse out. Normally the amount of water in the air around a leaf is less than that inside the leaf, so water diffuses to the lower water potential of the external air through the stomata whenever they are open. This process is called transpiration (see opposite), and it is a passive process, using energy from the sun, which evaporates water from mesophyll cells. As a result, vast amounts of water can be lost from a large tree. As water vapour diffuses from the air spaces in the mesophyll and through the stomata, it is replaced by water from the mesophyll cells. This in turn is replaced by water from the xylem in the veins of the leaf.

Since the xylem is a continuous system of tubes, water is drawn through to replace the water lost from the uppermost ends of the xylem, a little like the way a drink moves up a straw into your mouth when you suck. However, since water cannot be pulled up more than 10 m, even the most fantastic 'sucker' cannot drink through a straw longer than 10 m. Trees are often much taller than this, so how do they overcome this problem?

The answer is that xylem vessels are very narrow, and water molecules tend to stick together because of the hydrogen bonds between them. This is why water droplets form at the end of a tap: the water does not fall off until the weight is greater than the force holding the molecules together. This property of water is called **cohesion** (see opposite). As water moves out of the xylem in a leaf, it drags other molecules of water behind it. Because the vessels are so narrow, the column of water behind does not break, and water is pulled up all the way from the roots. The pulling force is so great that the column of water is actually being stretched. It is under **tension**, just as an elastic cord being pulled up would be. The tension in the column of water tends to pull the walls of the vessels inwards slightly. However, the lignin in the walls is strong enough to stop the vessels collapsing, just as the cartilages in the trachea and bronchi prevent them from collapsing as we expand the chest to breathe in (see page 118).

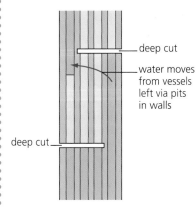

**Figure 10.2** Diagram to illustrate the cut stem experiment.

## Extension

### Vessel damage

Surprisingly, the diameter of a tree trunk is reduced very slightly, but measurably, when the tree is transpiring rapidly on a hot day, because the tension in the xylem is sufficient to pull in the walls of the many vessels just a little. If the column of water in a vessel is broken, for example by an air bubble or a cut, the water will ping apart like elastic, leaving an empty section above and below the bubble or cut. This is why pits are important. They permit water to move from one vessel to a neighbour if a vessel is damaged. A simple experiment showed this. The stem of a young tree was cut over half way across at two positions, one above the other, as seen in Figure 10.2. This ensured that all vessels were cut. Even so, water continued to rise up the stem by moving sideways between the cuts.

The two ideas, of cohesion and tension, used to explain how water is pulled up in a plant (Figure 10.3), are brought together in the **cohesion–tension theory**. The theory explains how water reaches the leaves of a plant from its roots.

## Transpiration

Water enters the roots by osmosis and passes up the stem to the leaves, where it evaporates into the air spaces inside the leaf and then passes out to the atmosphere. The air has a low water potential because it normally has a low percentage of water vapour.

Most of the water taken in through the roots of plants living in damp climates simply passes up through the xylem and then out from the leaves. Only a small proportion is used in photosynthesis to manufacture glucose. The loss of so much water by transpiration may seem very wasteful, but it is the unavoidable effect of the need for leaves to take in carbon dioxide for gas exchange. It does, however, have some advantages. The stream of water also transports mineral ions around the plant. The evaporation of water from the leaves has a cooling effect, just as the evaporation of sweat from our skin does. When leaves are exposed to bright sunlight, transpiration can reduce the possibility of the leaves overheating and the enzymes being denatured.

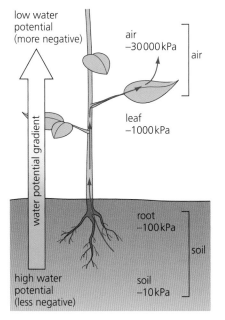

**Figure 10.3** The passage of water through a plant.

Polar molecule A molecule with a slight positive charge in one part of the molecule and a slight negative charge in another part.

## The properties of water

The movement of water through xylem vessels and the cooling effect of transpiration both depend on the properties of water. Water molecules are polar molecules because the hydrogen atoms have a slight positive charge and the oxygen atom has a slight negative charge. Opposite charges attract one another so water has strong **cohesion** between the molecules (see Figure 10.4).

**Figure 10.4** Water molecules form hydrogen bonds between molecules, so they 'stick' together.

slight negative charge

Electrons are pulled towards the oxygen atom, so the oxygen atom has a slight negative charge and the hydrogen atoms a slight positive charge.

slight positive charge

The slight charges cause an attraction between water molecules, which is called cohesion.

**TIP**

These properties of water are important in many parts of the course, and you may need to apply these ideas to questions in other topics.

**Figure 10.5** Pond skaters are predators that can move around by using surface tension.

**Latent heat of vaporisation** The heat required to turn liquid at its boiling temperature into gas at the same temperature.

**Heat capacity** The heat required to raise the temperature of a volume of liquid by 1°C.

The fact that water molecules are drawn towards one another explains why columns of water can be pulled up xylem vessels. The height of the tallest trees is probably limited by the cohesiveness of water. Coastal redwoods (*Sequoia sempervirens*) grow to over 90 m tall. Much taller than this and the long columns of water in their xylem vessels would snap too often to allow water to reach the highest branches.

This property of water also explains why surface tension exists. It is exploited by small specialised animals such as pond skaters, which can move around on the surface of ponds and ditches. Water molecules at the surface are attracted more strongly to one another than those that are completely surrounded by other water molecules. This is because where water and air meet there are no water molecules above them. The stronger cohesion between the layer of surface molecules creates a sort of film on which animals such as pond skaters can stand (see Figure 10.5).

The cooling effect of transpiration that protects leaves from overheating is due to water's relatively large **latent heat of vaporisation**. The relatively large latent heat of evaporation is due to the fact that water molecules are cohesive and so relatively difficult to separate. Even a small amount of evaporation into the leaf air spaces cools the leaf. This can be critical for a plant's survival in direct sunlight. The same process takes place to cool down an animal's body when sweat evaporates from the skin.

Water has other properties that have biological significance. It has a relatively high **heat capacity** so large volumes of water such as ponds or lakes do not change temperature very quickly. This resistance to temperature change means that aquatic organisms such as fish live in a very stable habitat. In particular, fish that live in small rock pools on tidal shores depend on this property of water for their survival. Due to the high heat capacity of water, small rock pools do not heat up too much in direct sunlight, protecting the organisms that live there. This also helps larger organisms, which are mainly composed of water, to maintain a stable temperature.

Water is also the **solvent** in which biological reactions take place in the cytoplasm of cells and in mass transport fluids. A wide range of substances are soluble in water, which is why so many different solutes can be transported in blood, lymph and plant sap. Water also takes part in many metabolic reactions itself, including hydrolysis and condensation reactions (see Chapter 1, page 2).

## TEST YOURSELF

**1** List four features of xylem vessels that make them specialised for the transport of water.
**2** What property of water molecules does the upward transport of water depend on? Explain your answer.
**3** Give two advantages for plants in losing so much water through their leaves.
**4** Suggest the combination of environmental factors that would lead to the most rapid rate of transpiration.

## Translocation

Water is not the only substance moved by mass transport in plants. A separate transport system moves organic substances around plants. The process is called **translocation** and, unlike xylem vessels, which are tubes made of dead cells linked end to end, it depends on the activity of chains of living cells, called **phloem sieve tubes**.

Like xylem vessels, phloem sieve tubes are elongated cells joined end to end to form a chain, as shown in Figure 10.6a. Where the cells meet, their end walls have holes, which is why they are called sieve plates. The chains of cells run parallel to one another in bundles in the stem, close to similar bundles of xylem vessels. Bundles of phloem sieve tubes in a stem are shown in Figure 10.6b.

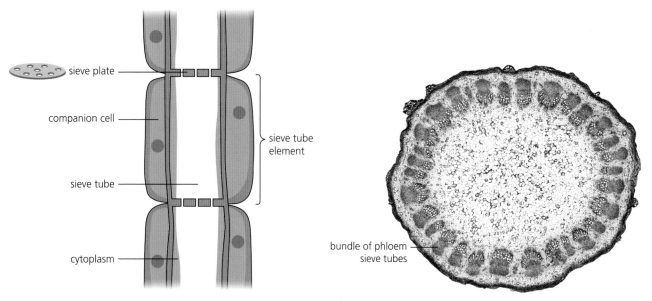

**Figure 10.6** (a) Phloem sieve tubes are made of elongated cells joined end to end at sieve plates.

(b) Cross-section of a stem showing phloem tissue.

Each separate cell in a phloem sieve tube is called a **sieve element** and has another cell alongside it called a **companion cell**. Companion cells provide metabolic support to the sieve elements. This is important because the cells of sieve tubes lose many of their organelles, including their nucleus, as they become specialised. Although this allows easier flow of phloem sap through the cell, it means that they cannot repair and maintain themselves so well and rely instead on their companion cells for many of these functions.

One of the most important substances translocated in phloem is the sugar sucrose. Sucrose is made in leaves from the products of photosynthesis. It is translocated to parts of the plant that are growing or to places where sucrose can be stored, often used to make starch. Leaves and storage organs such as potato tubers are known as **sources** because they produce sucrose, whereas buds, developing seeds and other sucrose-using parts of the plant are called **sinks**. Translocation occurs from sources to sinks.

**Sap** The fluid that is transported in phloem.

Mass transport in phloem is explained by the mass flow hypothesis. Evidence suggests that phloem sieve tubes are under pressure rather than under tension. This is why phloem sap leaks out if sieve tubes are punctured. The mass flow hypothesis suggests that phloem sap is pushed from source to sink by a **hydrostatic pressure** gradient caused by osmosis.

Sucrose molecules are too large to simply diffuse across membranes. In the veins of leaves, companion cells known as **transfer cells** actively load sucrose into the phloem against a concentration gradient. Figure 10.7 shows that the loading is similar to that you saw for glucose uptake in the epithelial cells of the intestine (see Chapter 8, page 143). In this case, hydrogen ions rather than sodium ions are actively pumped out of the transfer cell, creating a hydrogen ion gradient. The sucrose is transported in along with hydrogen ions by a co-transport protein and diffuses into the neighbouring sieve element.

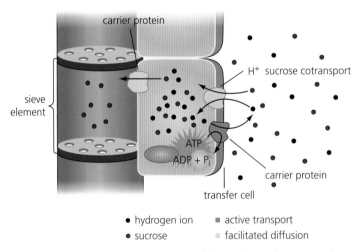
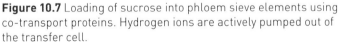

**Figure 10.7** Loading of sucrose into phloem sieve elements using co-transport proteins. Hydrogen ions are actively pumped out of the transfer cell.

As sucrose accumulates in the sieve element, the water potential is lowered, so water moves into the sieve tube down a water potential gradient by osmosis from nearby xylem vessels, increasing the hydrostatic pressure. Because the opposite process happens where sucrose is unloaded from phloem, water leaves the sieve tubes by osmosis at sinks, lowering the hydrostatic pressure. The pressure gradient between source and sinks causes the mass flow of phloem sap in one direction along the sieve tube. The mechanism of the mass flow hypothesis is outlined in Figure 10.8.

**Figure 10.8** The mechanism of the mass flow hypothesis.

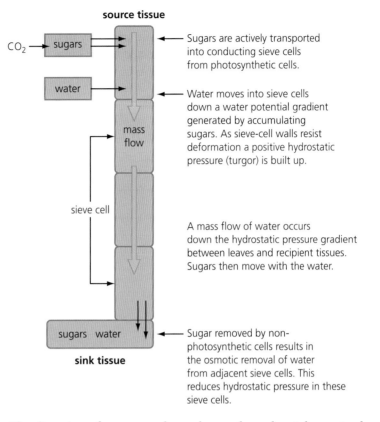

The direction of transport depends on where the sinks are in the plant at any point in time. When growth is taking place, areas undergoing cell division are the main sinks, so sap flows from the leaves both upwards to leaf buds and downwards to growing root tips. When a plant is using materials from a storage organ such as a potato tuber, the sap flows from the storage organ to the buds. A variety of organic substances other than sucrose are translocated by the phloem, including amino acids, plant growth hormones and messenger RNA. Viruses can also be transported in the phloem.

## Evaluating the evidence for the mass flow hypothesis

Some simple observations support the mass flow hypothesis. If a phloem sieve tube is punctured then phloem sap oozes out, suggesting it is under pressure. Phloem sap sampled from a source has a higher sucrose concentration than sap sampled from a sink, confirming that different water potentials would cause osmosis into or out of the sieve tubes in the two locations. If a plant virus is applied to well-illuminated leaves, viruses can be detected moving downwards in the phloem towards the roots. But if the virus is applied to leaves in the dark, they are not transported, suggesting that the production of sucrose by photosynthesis is required at the source for translocation to occur.

On the other hand, some evidence contradicts the mass flow hypothesis. Measurements of the rate of translocation of different organic substances suggest that amino acids travel more slowly than sucrose. Some scientists claim to have detected different substances moving in opposite directions in the same sieve tube. If the mass flow hypothesis were correct, bulk flow of phloem sap ought to occur in the same direction at the same rate in any one sieve tube, and probably in most sieve tubes at the same time because they are all connected to the same source: the leaves. These observations suggest a more complex mechanism may be responsible for translocation.

> **TIP**
> When you evaluate any data or evidence, you should always consider both sides of the argument. In this case, there are observations that both support and contradict the mass flow hypothesis.

## ACTIVITY

### Investigating translocation in phloem

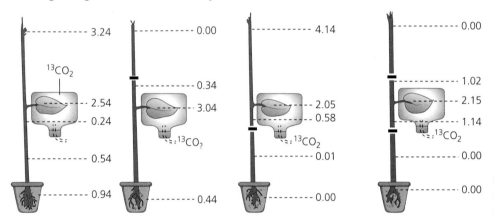

**Figure 10.9** Results of a ringing experiment using radioactive carbon dioxide.

Figure 10.9 shows the results of some experiments carried out to investigate translocation in phloem tissue. This type of experiment is called a **ringing** experiment, because it involves the removal of a ring of surface tissues from a plant stem (shown on the diagram by small black bars dividing the stem), leaving the core of the stem intact. You will recall the cross-section of a plant stem from Figure 10.1. The rings are cut deeply enough to remove the phloem, but they leave the xylem intact in the core of the stem. The plant can then be supplied with a radioactive **tracer** to investigate the direction and rate of translocation. Numbers on the diagram indicate the amount of radioactive carbon detected in each part of the plant.

1 Plants in this type of experiment are supplied with radioactive carbon dioxide. Explain how this works as a 'tracer' to detect the translocation of organic substances.
2 When removing a ring of material from a plant stem, why is it important to ensure that the xylem remains intact?
3 What is the purpose of the plant with no rings of tissue removed?
4 How do the results of the ringed plants support the mass flow hypothesis?

Another way to use radioactive **tracers** is to make images of the plant showing where radioactive substances have been transported. After allowing the plant to take up radioactive carbon dioxide and carry out photosynthesis and translocation, the plant is pressed against photographic paper in the dark for several hours. Any radioactive material in the plant tissues creates an image called an **autoradiograph**.

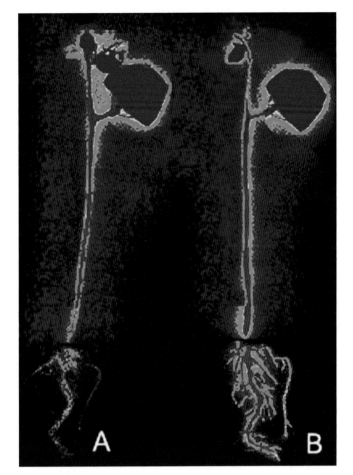

**Figure 10.10** Radioactive carbon dioxide used as a tracer to detect the translocation of organic substances from the leaves to the roots.

Autoradiographs taken at a series of times after exposure to radioactive carbon dioxide can be used to calculate how long it takes for sucrose and other organic products of photosynthesis to travel from the leaves to other parts of the plant such as roots, flowers or fruits.

## TEST YOURSELF
**5** Suggest why transfer cells have numerous mitochondria.
**6** Potatoes store sucrose as starch. Describe how the mass flow hypothesis could explain sucrose translocation from leaves to potatoes as they grow.
**7** Rabbits and deer often eat the bark of young trees and the phloem tissue beneath it in winter. If they eat all round the tree trunk, this can kill the tree. Suggest why.
**8** Plant growth hormone sprayed onto the leaves of a plant is absorbed into the phloem sap. Used as a weedkiller, it can kill the whole plant, including the roots. Suggest how.

# Practice questions

**1 a)** Give one feature of a plant cell that enables you to identify it as a xylem vessel. *(1)*

The figure below shows a potometer.

**b)** Explain how you could use this apparatus to find the volume of water taken up by this shoot in a minute. *(3)*

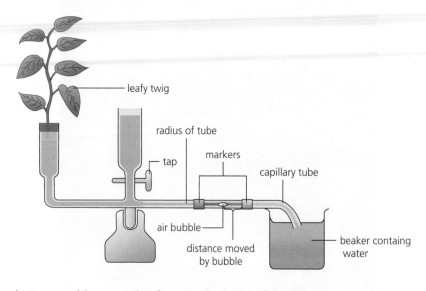

leafy twig

radius of tube

markers

tap

capillary tube

air bubble

distance moved by bubble

beaker containg water

**c)** You would expect this figure to be lower if the shoot was placed in humid conditions. Explain why. *(2)*

**d)** If the diameter of the capillary in this potometer were 0.6 mm and the bubble moved 37 mm in 12 minutes, calculate the rate of water loss by the leafy shoot. *(1)*

**2** A group of students carried out an investigation. They filled 40 test tubes with tap water until the water was about 1 cm from the top of the tube. They carefully marked the level of the water in the tube by using a pen on the outside of the tube. They covered each tube with a small piece of clingfilm.

Next, they obtained 40 fresh leaves from a plant, checking that all the leaves were similar in size. They inserted one leaf into each test tube, making a small hole in the clingfilm so that the leaves were all in the water. They divided the leaves into four groups of 10. They applied petroleum jelly (a water- and gas-tight substance) to some of the leaves, as described in the table.

| Group | Treatment |
|-------|-----------|
| A | Petroleum jelly applied to both surfaces of the leaf |
| B | Petroleum jelly applied to upper surface of leaf only |
| C | Petroleum jelly applied to underside of leaf only |
| D | No petroleum jelly applied at all |

The test tubes were all placed in test tube racks on a laboratory bench, away from bright sunlight or draughts. They were left for a week. The figure shows one of the tubes after a week.

leaf

clingfilm

level of water

test tube

After a week, the students removed the leaves and clingfilm from the tubes. They carefully measured the volume of water needed to bring the level of water in the tubes back up to the level marked with the pen. They found the mean volume of water needed by the tubes in each group. The results are shown in the table.

| Group | Treatment | Mean volume of water added/cm³ |
|---|---|---|
| A | Petroleum jelly applied to both surfaces of the leaf | 3.2 |
| B | Petroleum jelly applied to upper surface of leaf only | 4.8 |
| C | Petroleum jelly applied to underside of leaf only | 3.7 |
| D | No petroleum jelly applied at all | 5.1 |

**a) i)** Explain why each test tube had a lid of clingfilm. *(1)*

**ii)** Explain why it was important to have 10 leaves in each group. *(2)*

**b) i)** What conclusions could be drawn from these results? Use your knowledge of leaf structure to explain the results. *(4)*

**ii)** Explain the purpose of group D in this investigation. *(2)*

**3** In an investigation, a scientist used a very accurate measuring device to measure the diameter of a tree trunk over a period of several days. Some of the results obtained are shown in the figure.

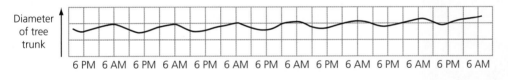

Diameter of tree trunk

6 PM  6 AM  6 PM  6 AM  6 PM  6 AM  6 PM  6 AM  6 PM  6 AM  6 PM  6 AM  6 PM  6 AM

**a) i)** Describe these results. *(3)*

**ii)** Use your knowledge of water transport through plants to explain these results. *(4)*

In a different investigation, a scientist set up the apparatus shown in the figure below.

**b) i)** Parts X and Y of this apparatus represent different parts of a plant. Name the parts of a plant that they represent. *(2)*

**ii)** As water evaporates from X, the mercury at Y rises up the tube. The scientist said that this is evidence to support the cohesion–tension theory of water transport in plants. Do you agree with this? Give reasons for your answer. *(4)*

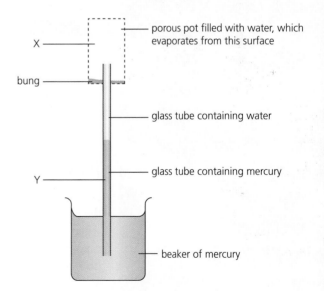

porous pot filled with water, which evaporates from this surface

X

bung

glass tube containing water

glass tube containing mercury

Y

beaker of mercury

## Stretch and challenge question

**4** The mechanism of phloem transport remains controversial amongst scientists. In particular, recent criticisms suggest that the mass flow hypothesis does not seem to be adequate to explain translocation in tall trees. Explain why.

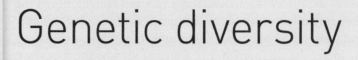

# Genetic diversity

**PRIOR KNOWLEDGE**

*Before you start, make sure that you are confident in your knowledge and understanding of the following points.*

● Different genes control the development of different characteristics of an organism.
● Differences in the characteristics of different individuals of the same kind may be due to differences in the genes they have inherited (genetic causes), the conditions in which they have developed (environmental causes) or a combination of both.
● Particular genes or accidental changes in the genes of plants or animals may give them characteristics which enable them to survive better.
● Evolution occurs by natural selection.
● Individual organisms of a particular species may show a wide range of variation because of differences in their genes.
● Individuals with characteristics most suited to their environment are more likely to survive to breed successfully.
● The genes that have enabled these individuals to survive are then passed on to the next generation.
● Where new forms of a gene result from mutation there may be relatively rapid change in a species if their environment changes.
● Many strains of bacteria have developed resistance to antibiotics as a result of natural selection.

**TEST YOURSELF ON PRIOR KNOWLEDGE**

**1** Give two reasons why organisms in a population may look different from one another.
**2** What is a mutation?
**3** Why is the appearance of antibiotic resistance in bacteria such a problem?

## Introduction

Scientists analysed DNA samples from the skins of tigers in the collection at the Natural History Museum in London, dating from 1858 to 1947, a period when many tigers were hunted and their skins preserved. They found that the mitochondrial DNA of modern tigers has only 7% of the genetic diversity of historic tigers. Large populations of tigers once existed, but as a result of hunting and habitat loss they have been fragmented into much smaller populations. This means that the number of alleles in the gene pool of each small population is smaller than in the original population. Since they do not exchange alleles by mating with tigers from other populations, their genetic diversity is likely to remain very low. Each time another small population becomes locally extinct, their alleles are lost. A low genetic diversity may prevent the remaining tiger populations from adapting to a changing environment and may threaten their long-term survival just as much as further loss of their habitat.

Future conservation efforts will need to concentrate on trying to connect patches of forest with habitat through which tigers are able to move as well as protecting the patches themselves. Allowing tigers to mix and mate with individuals from other small populations will improve the connectivity of the gene pool and help to maintain what is left of their genetic diversity.

You have seen how the genetic diversity of a species can be reduced if populations become fragmented and the number of organisms in the smaller populations falls markedly. Alleles can be lost from the gene pool and the ability of the population to adapt to change is limited. On the other hand, genetic diversity is increased by random mutation, which results in new alleles being added to the gene pool of the population. Genetic diversity can also be increased by chromosome mutations and by random factors associated with meiosis and fertilisation. Genetic diversity is necessary for natural selection to take place and is crucial in enabling a population to adapt to a changing environment.

**Allele** Alleles are different forms of the same gene.

**Genetic diversity** A measure of the number of different alleles of genes in a population.

**Gene pool** All of the different alleles of genes in a population.

**Population** A group of organisms belonging to the same species found in the same area at the same time and potentially able to interbreed.

# Meiosis

Look at Figure 11.1. It shows the chromosomes from one cell of a human female. To make this, a photograph was taken of the cell using a camera and an optical microscope. The image of each chromosome was then cut out of the photograph and the images were arranged as you see them. Notice that they have been arranged in pairs. The members of each pair are the same size and shape. More importantly, they carry genes controlling the same characters in the same order. We call the members of each pair **homologous chromosomes**. You inherited one member of each homologous pair of chromosomes from your mother (**maternal chromosome**) and the other member of the homologous pair from your father (**paternal chromosome**). The fusion of these gametes in sexual reproduction produces genetic diversity in a population because each parent may contribute different alleles of each gene.

A cell with pairs of homologous chromosomes is called diploid. A cell with only one chromosome from each homologous pair is called haploid. We often refer to diploid cells as $2n$ and haploid cells as $n$. In Figure 11.1 you can see that diploid human cells have 46 chromosomes ($2n = 46$). However, not all our cells are diploid. Our gametes – the egg and sperm cells that we produce – are haploid. Human gametes (the sex cells) have 23 chromosomes ($n = 23$).

The type of cell division that produces haploid cells from diploid cells is called **meiosis**. Most eukaryotic organisms have diploid and haploid cells during their life cycle. This means that meiosis occurs at some stage in the life cycle. Figure 11.2 (overleaf) shows two life cycles. Notice that, although humans produce gametes by meiosis, some organisms such as fungi do not.

Meiosis has one other important effect: the haploid cells produced are genetically different from each other. There are two reasons for this: independent assortment of homologous chromosomes and genetic recombination by crossing over.

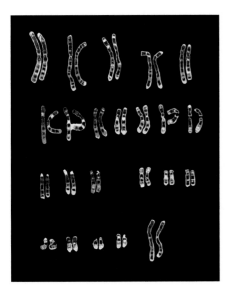

**Figure 11.1** The chromosomes in a body cell from a human female.

## Independent assortment of homologous chromosomes

The heading might sound a complicated description of the process, but the sequence is quite easy to follow. Figure 11.3 (overleaf) shows a cell with two pairs of homologous chromosomes. The members of each pair are colour-coded to distinguish the maternal and paternal chromosomes. Before meiosis starts, each chromosome makes a copy of itself by DNA replication. These copies are now called chromatids.

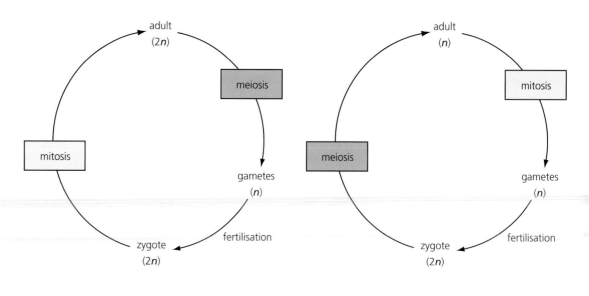

**(a)** Human life cycle

**(b)** Life cycle of fungus

**Figure 11.2** Meiosis produces haploid cells (*n* chromosomes) from diploid cells (2*n* chromosomes). (a) In humans, meiosis occurs during gamete production. (b) In fungi, meiosis occurs after fertilisation, not before it.

**Figure 11.3** The result of meiosis in a cell with two pairs of homologous chromosomes. Before meiosis occurs, the chromosomes were copied by DNA replication. As a result of meiosis, four cells (haploid gametes) can be formed, and there are two possible combinations. Another cell with the same two pairs of homologous chromosomes can produce another two different combinations (AB and ab), making four possible combinations in total.

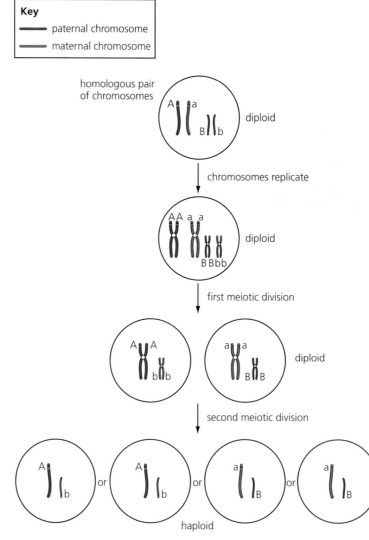

Then two divisions occur:

- the **first meiotic division** separates the chromosomes in each homologous pair
- the **second meiotic division** separates the two chromatids made by DNA replication before meiosis started.

In the first meiotic division, the separation of one pair of homologous chromosomes is not affected by the separation of any other pair. In other words, if the maternal chromosome from one pair moves to one side of the cell, it does not cause the maternal chromosomes of other pairs to go in the same direction. This is independent assortment of chromosomes. The result is that gametes contain a mixture of maternal and paternal chromosomes.

Remember that homologous chromosomes carry genes controlling the same characters in the same order. However, each gene can have different forms, called alleles. To take a simple example, the human gene for haemoglobin has two forms; one leads to sickle cell anaemia and one does not. The maternal and paternal chromosomes in one homologous pair carry different alleles of many of their genes. Therefore, independent assortment produces different combinations of alleles in the cells formed during meiosis. If you look back to Figure 11.3, you will see that a cell with just two pairs of homologous chromosomes could produce haploid cells with any one of four different combinations of maternal and paternal chromosomes. As you know, human cells have 23 pairs of homologous chromosomes. Independent assortment of 23 pairs of homologous chromosomes can result in $2^{23}$ different chromosome combinations (see page 245) in the haploid egg cells and sperm cells.

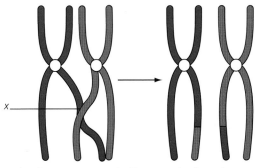

breakage occurs at point X; fragments join 'wrong' chromosome

result of crossing over

**Figure 11.4** During meiosis, if the chromatids of a homologous pair break at the same point, DNA may be exchanged between the chromatids in opposite members of a pair. This is called crossing over.

## Genetic recombination by crossing over

The chromosomes at the end of meiosis in Figure 11.3 are the same colour code as they were at the beginning. Figure 11.4 shows how homologous chromosomes can exchange pieces of their DNA.

During the first meiotic division, the members of each homologous pair lie side by side. If chromatids become tangled with one another, they may break and the broken segments may be rejoined to chromatids in opposite members of the pair. This is called **crossing over** and results in **recombination** or new combinations of alleles that gives rise to genetic diversity.

The effect of crossing over is shown in Figure 11.5 (overleaf). You can see in this diagram that some of the haploid cells have chromosomes with part of the DNA from both the maternal and paternal chromosomes. They have the same genes but a combination of alleles that was not present in either of the parental chromosomes.

195

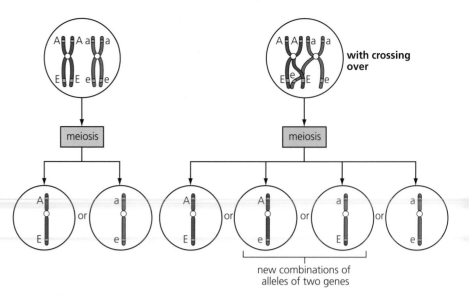

**with crossing over**

new combinations of
alleles of two genes

**Figure 11.5** Crossing over results in cells with chromosomes that carry new combinations of alleles.

**Chromosome mutation** A change in the number of chromosomes.

If chromosomes fail to separate properly during meiosis, then a gamete may end up with two copies of a chromosome rather than just one. This is called **non-disjunction**. That gamete would contain one more than the haploid number and, should it take part in fertilisation, it would result in a chromosome mutation, where the organism would have the wrong number of chromosomes in its diploid cells (Figure 11.6).

**Figure 11.6** Human chromosomes showing the chromosome mutation that causes Down's syndrome.

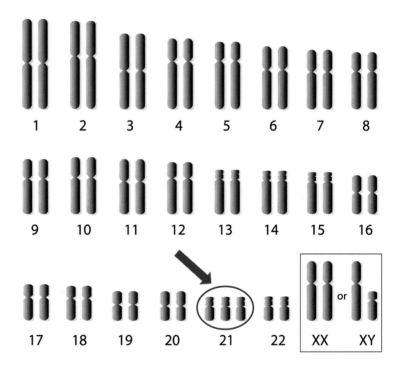

**TEST YOURSELF**

1 We often use the symbol 2*n* to represent diploid cells. What does *n* represent?
2 Explain one advantage to humans of producing haploid gametes.
3 Use your knowledge of independent assortment of homologous chromosomes to explain why a human couple could have children who are all genetically different from each other.

base sequence in DNA →→→ processes in cell →→→ amino acid sequence in polypeptide

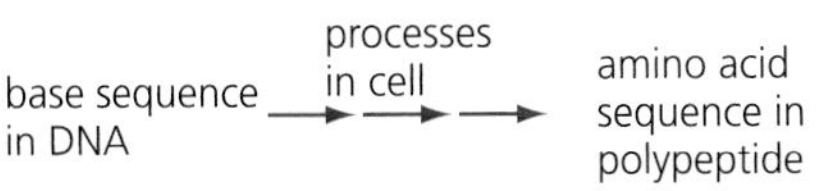

**Figure 11.7** A polypeptide is determined by the genetic code held in the DNA base sequence of a gene.

Gene mutation A change in the base sequence of a gene.

# Gene mutations

You learned in Chapter 2 that enzymes control reactions in cells and that enzymes are protein molecules. You learned in Chapter 4 that a gene is a base sequence of DNA that codes for the sequence of amino acids in a particular polypeptide, or protein. Figure 11.7 summarises the relationship between DNA and proteins. Although their cell structures are different, prokaryotic cells and eukaryotic cells use their genes to make polypeptides in a similar way.

A change in the DNA base sequence of a gene is called a gene mutation, forming a new version of the gene, or allele. A gene mutation generally happens when DNA is replicated. You learned about DNA replication in Chapter 5. A gene mutation is rare, occurring about once every 1 million times that a gene is copied, and is a completely random and spontaneous event. Sometimes a gene mutation causes the encoded protein to lose its function as an enzyme.

A gene mutation often causes a change in the sequence of amino acids in the encoded protein. Often this has no effect on the function of that protein. These mutations are called **neutral mutations** and are probably the most common type of gene mutation. However, some gene mutations cause a change in the amino acid sequence of the encoded protein, and its function is lost. As a result, these mutations are harmful. It is extremely rare for a gene mutation to produce a beneficial change in the activity of the encoded protein. If it does, it means that the individual in which the mutation has occurred has some sort of advantage over other members of the population. This can lead to the mutation being passed on, or inherited, by its offspring.

Two types of gene mutation that may occur are outlined here.

- **Base deletion**: a base is lost from the base sequence. As a result, the whole base sequence following the deleted base moves back one place (a **frame shift**). This type of mutation often has a significant effect on the encoded protein because it can alter the sequence of all of the codons following the lost base.
- **Base substitution**: the 'wrong' base is included in the base sequence. It does not cause a frame shift but might result in a different amino acid being included in the polypeptide chain. On the other hand, if the substitution results in a triplet that still codes for the same amino acid, it may not change the sequence of amino acids at all because the genetic code is degenerate (see Chapter 4, page 70).

Table 11.1 shows the effects that can arise from base deletion and base substitution.

**Table 11.1** The first two rows of this table show the amino acid sequence encoded by part of a molecule of mRNA. The mRNA sequence is shown as individual codons to help you to read the code.

| Original base sequence on mRNA | AGA | UAC | GCA | CAC | AUG | CGC |
|---|---|---|---|---|---|---|
| Encoded sequence of amino acids | Arg | Tyr | Ala | His | Met | Arg |
| mRNA base sequence after base deletion on DNA | AGU | ACG | CAC | ACA | UGC | GCx |
| Encoded sequence of amino acids | Ser | Thr | His | Thr | Cys | Ala |
| mRNA base sequence after base substitution on DNA | AGU | UAC | GCA | CAC | AUG | CGC |
| Encoded sequence of amino acids | Ser | Tyr | Ala | His | Met | Arg |

## Mutagenic agents

Natural mechanisms occur within cells that identify and repair damage to DNA. These mechanisms become ineffective if the rate of mutation increases above the normal, low rate. Many environmental factors increase the rate of mutation. They are called **mutagenic agents** and include:

- toxic chemicals, for example bromine compounds and peroxides
- ionising radiation, for example gamma rays and X rays
- high-energy radiation, for example ultraviolet light.

# Natural selection

## A change in the frequency of alleles in a population

When a gene mutates, it has two forms: the original form and the mutated form. We call the different forms of the same gene alleles. If one of the alleles of a gene confers an advantage over the alternative allele of the same gene, it will probably become more common (its frequency will rise) in the population. We call this process **natural selection**, and the allele is known as an **advantageous allele**.

Natural selection can change the frequency of alleles in a population of a species, including in populations of humans.

Figure 11.8 shows two adults of a species of snail called *Cepaea nemoralis*, the banded snail. One of the snails has a yellow shell and the other has a pink shell. The difference in shell colour is controlled by two alleles of the gene for shell colour. The frequency of the alleles for yellow and pink shells is different in populations in different habitats. Table 11.2 summarises how natural selection affects the frequency of these alleles in two different populations. Notice that while natural selection acts on individuals, it leads to change in the gene pool of the whole population.

> **TIP**
> When talking about natural selection, you should refer to alleles rather than genes. The gene is for the character, such as shell colour, whereas alleles are the different forms of the gene that control the pink and yellow colours.

**Figure 11.8** The banded snail, *Cepaea nemoralis*, is common throughout Europe. The difference in shell colour is controlled by two alleles of a single gene.

**Table 11.2** Natural selection leads to differences in the frequency of two alleles controlling shell colour in populations of the banded snail, *Cepaea nemoralis*.

| Habitat in which snail population lives | More frequent allele in population | How natural selection affects frequency of alleles |
|---|---|---|
| Beech woodland in England | Pink allele | Snails are eaten by song thrushes, which can distinguish colour. Pink shells are camouflaged amongst the leaf litter but the yellow shells are conspicuous. The song thrushes eat more yellow-shelled snails than pink-shelled snails. |
| Grassland of the Pyrenees (a mountain range dividing France and Spain) | Yellow allele | No song thrushes live in these mountains. The yellow shells reflect heat from strong sunlight. The pink shells absorb more heat and the snails are more likely to die. |

## Increased reproductive success

Allele frequencies in a large population generally remain stable from generation to generation. This will not be true if some organisms are:

- more likely to survive until they reproduce
- more likely to grow sufficiently well to reproduce successfully
- more likely to attract a mate.

If the environment changes or a new advantageous allele arises by mutation, some organisms will tend to reproduce more successfully than others and will leave more offspring. We say that they have **increased reproductive success** compared with other individuals in the population.

In beech woodlands, the yellow-shelled snails shown in Figure 11.8 are very conspicuous against the pink leaf litter lying on the ground. The snails with pink shells are better camouflaged and so are more difficult to find. Song thrushes eat banded snails. Like us, song thrushes have colour vision. A large number of investigations have shown that song thrushes find more of the conspicuous yellow-shelled snails than they do pink-shelled snails in beech woodlands. As a result, fewer yellow-shelled snails survive to reproduce. This means that fewer of the alleles for yellow shells are passed on to the next generation. Table 11.3 summarises the process of natural selection.

> **TIP**
> The sequence of events on the left-hand column of Table 11.3 can be applied to any example of natural selection. You just need to tailor your explanation by adding details like those in the right-hand column.

**Table 11.3** An explanation of natural selection in a beech woodland population of banded snails. The events in the left-hand column can be applied to any example of natural selection.

| Sequence of events leading to natural selection | Application of these events to selection of yellow banded snails in beech woodlands |
|---|---|
| Within a population, a gene has more than one allele. This genetic diversity is due to random mutation. | The two alleles of the shell colour gene result in snails with pink shells and snails with yellow shells. |
| There is differential reproductive success between the organisms with different alleles of the same gene. | Yellow-shelled snails are more conspicuous than pink-shelled snails among beech litter. Song thrushes find yellow-shelled snails more easily than they find pink-shelled snails. Fewer yellow-shelled snails survive to reproduce. |
| Organisms with greater reproductive success leave more offspring than those with less reproductive success. | In a beech woodland, pink-shelled snails have more offspring than yellow-shelled snails. |
| Organisms with greater reproductive success will pass their advantageous allele to their offspring. As a result, the frequency of this allele will increase in the population, i.e. natural selection has occurred. | In a beech woodland, the frequency of the pink allele is higher and the frequency of the yellow allele is lower. Pink-shelled snails are at a selective advantage in beech woodlands. |

> ## TEST YOURSELF
>
> **4** A gene mutation involves a change in the base sequence of a gene. Where in the DNA would a mutation *not* be likely to cause a change in the encoded protein?
>
> **5** Populations of banded snails also live in grassland, where they are preyed on by song thrushes. Which shell colour would you expect to be more common in grasslands? Explain your answer.
>
> **6** Explain what an advantageous allele is.
>
> **7** Why are advantageous mutations relatively rare?
>
> **8** Use the top two rows in Table 11.1 to explain how a base deletion changes the function of an encoded protein.
>
> **9** Use the bottom two rows in Table 11.1 to explain why not all base substitutions cause a change in the sequence of encoded amino acids.

# Different effects of natural selection

Natural selection operates on most characteristics that organisms possess and results in populations being better adapted to their environment. It can result in the frequency of an advantageous allele increasing in the gene pool (**directional selection**) or it can result in unfavourable alleles becoming rare in the gene pool (**stabilising selection**).

## Directional selection

If a gene mutation gives rise to two forms in a population, such antibiotic resistant and non-resistant bacteria, and there is a selection pressure acting on the population, such as the use of antibiotics, **directional selection** acts against one form and favours the other. Organisms with the advantageous allele are more likely to survive. As a result, one form becomes rare and the alternative form becomes common. Figure 11.9 shows how this happens for antibiotic resistance. The frequency of the resistance allele increases over generations.

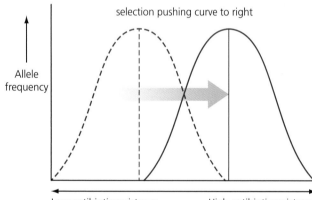

**Figure 11.9** Directional selection acts to increase antibiotic resistance.

## Stabilising selection

**Stabilising selection** acts against extreme phenotypes in a population. If some phenotypes are more optimal then the frequency of the alleles for these phenotypes remains the same while the frequencies of other forms that are somehow at a disadvantage are reduced. Stabilising selection occurs on birth mass in humans. Figure 11.10 shows that babies of birth mass within a particular range are at an advantage. Compared with those with alleles that result in very high or very low birth masses, babies with a more optimal birth mass have a lower infant mortality and so the frequencies of alleles for optimal birth masses tend to remain stable over generations.

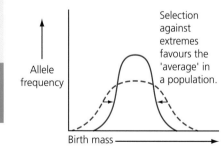

**Figure 11.10** Stabilising selection acts to maintain optimal human birth mass.

> ## TIP
> Directional selection usually happens when the environment changes, whereas stabilising selection usually happens when the environment remains the same.

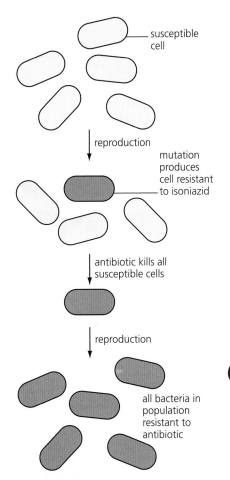

susceptible cell

reproduction

mutation produces cell resistant to isoniazid

antibiotic kills all susceptible cells

reproduction

all bacteria in population resistant to antibiotic

**Figure 11.11** Evolution of antibiotic-resistant populations of bacteria in the body of a TB patient. As a result of a gene mutation, a single bacterial cell becomes resistant to the antibiotic isoniazid, and this one cell survives the patient's treatment with the antibiotic. All the offspring of the cell inherit the allele for resistance, and soon the patient's whole population of TB bacteria is resistant to isoniazid.

## Natural selection and antibiotic resistance

The increase in antibiotic resistance is explained by natural selection acting on mutations for resistance that arise by chance in bacteria. This presents us with a major medical problem for the effective treatment of many bacterial diseases in the future. Some bacteria, such as those that cause tuberculosis (TB), have already evolved resistance to a range of antibiotics. Imagine a population of bacteria in the body of a TB patient. Figure 11.11 represents a small sample of these bacteria. Initially, none of the bacterial cells is resistant to the antibiotic isoniazid. By chance, a mutation happens to a gene in one bacterium. As a result, it becomes resistant to isoniazid. The patient is being treated with isoniazid, and so cells without the mutation – the susceptible cells – are killed by the antibiotic. Only the one cell carrying the gene mutation by chance – the resistant cell – survives. The patient now has just one bacterial cell. However, as bacteria can double in number in a matter of minutes, the patient soon becomes infected by a population of millions of isoniazid-resistant bacteria. There has been a fundamental change in the bacterial population; all the TB bacteria now carry an advantageous allele that gives a new characteristic beneficial to the bacterium.

**TIP**

You should refer to bacteria being *resistant* to antibiotics rather than *immune* to them. Immunity involves an immune response, antibodies rather than antibiotics and memory cells (remember Chapter 6). In this case, bacteria simply resist the effects of the antibiotic.

Table 11.4 shows a general example of how natural selection can change the frequency of alleles in a population of bacteria.

**Table 11.4** Populations of bacteria become resistant to antibiotics through natural selection.

| How natural selection explains antibiotic resistance in bacteria |
|---|
| A chance mutation results in an allele that gives resistance to an antibiotic to bacteria that possess it, so when the antibiotic is present they are at an advantage. |
| ↓ |
| In the presence of this antibiotic, bacteria with the resistance allele are able to survive and reproduce more successfully than those with the original allele. |
| ↓ |
| Bacteria with the resistance allele reproduce. |
| ↓ |
| As the resistance allele is passed on to more offspring, the frequency of the resistance allele will be greater in the population in the next generation than it was in the parental generation. |

**TIP**

As you can see in Table 11.4 the generic sequence of events for natural selection from Table 11.3 has been tailored to fit a bacterial resistance example.

## REQUIRED PRACTICAL 6

### Use of aseptic techniques to investigate the effect of antimicrobial substances on microbial growth

**Note: This is just one example of how you might tackle this required practical.**

Bacteria can be tested for their resistance to antibiotics using a method called disc diffusion. A sterile nutrient agar plate is prepared. The bacterium to be tested is cultured in a broth (see Chapter 15, page 263).

1 What is nutrient agar and why must it be sterile?
2 What is a broth and what does cultured mean?

After a period of incubation, the broth is then diluted to a standard concentration, which is normally $1 \times 10^8$ CFU mm$^{-3}$. CFU stands for colony-forming unit. In the case of bacteria, a colony-forming unit means a live bacterial cell capable of dividing and forming a colony on the agar.

The concentration of the diluted broth is given in standard form (see Chapter 14, page 244).

3 Write the number of live bacterial cells in each cubic millimetre of the diluted broth in ordinary form.

A sample of the diluted broth is spread onto the surface of the sterile agar using aseptic technique. Paper discs pre-soaked in a standard concentration of each antibiotic are then pressed lightly onto the surface of the agar. It is important that the discs are spread out evenly and not too close to the sides of the plate or to each other.

4 Describe what is meant by aseptic technique and give examples of some of the steps that might be taken.
5 Why is it important that the concentrations of antibiotic are standardised?

The agar plate is then incubated overnight. During incubation, the antibiotic will diffuse outwards from each disc. This creates a gradient of antibiotic concentration.

The antibiotic is most concentrated nearest the disc. The further from the disc it diffuses, the more the antibiotic concentration decreases.

6 Suggest why the discs must be spread out carefully on the agar surface.

If the bacteria are susceptible to an antibiotic, a clear area is visible around the disc until a concentration is reached where the bacteria are no longer susceptible. The more effective the antibiotic, the lower the concentration at which it will kill the bacteria and the larger the clear zone. Figure 11.12 shows an example of a disc diffusion test. There are no bacteria in the clearer yellow areas.

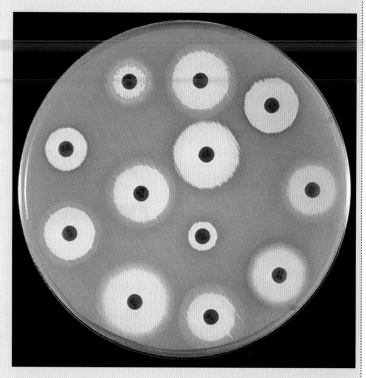

**Figure 11.12** Bacteria growing on a plate with antibiotic discs.

7 What causes the clear zones?
8 The species of bacteria growing on this plate is not completely resistant to any of the antibiotics. Explain how you know this.
9 How could you tell which antibiotic is least effective against this species of bacteria? Explain your answer.
10 If you wanted to assess the effectiveness of these antibiotics against this species of bacterium, you could compare the radius of each clear zone. However, measuring the diameter would be easier. Suggest why.
11 What problems do you think you might have in making these measurements?
12 Draw a suitable results table for collecting the data from this plate. The Petri dish in Figure 11.12 is life size. Use a ruler to measure the clear zones and record and present your measurements in an appropriate way (see Chapter 14, page 240).
13 Which antibiotic do you think is most effective?

# Extension

## Variation can be caused by environmental factors as well as by genetic factors

Tall people tend to have tall children. This suggests that height is inherited. However, children born after the introduction of the National Health Service in 1948 grew to be taller, on average, than those born before 1948. It is unlikely that there had been a genetic change in the population. It is more likely that children grew taller because they had a better diet and were free from disease. These are environmental, not genetic, factors.

How easy is it to tell whether variation is caused by genetic or environmental factors? A student measured the height of the students in year 1 of A-level biology. Figure 11.13 summarises her results. What can we conclude from these histograms? You can see that there is no clear-cut 'tall' or 'short' person. The range of heights shows continuous variation (see Chapter 14, page 252). You can see that the distribution of heights is different for males and females. This suggests that these differences result from genetic differences between females and males. You can also see that there is a range of values around the peak values. Is this caused by genetic factors or by environmental factors? Without further information, we cannot easily draw conclusions about which of these two types of factor is the more important.

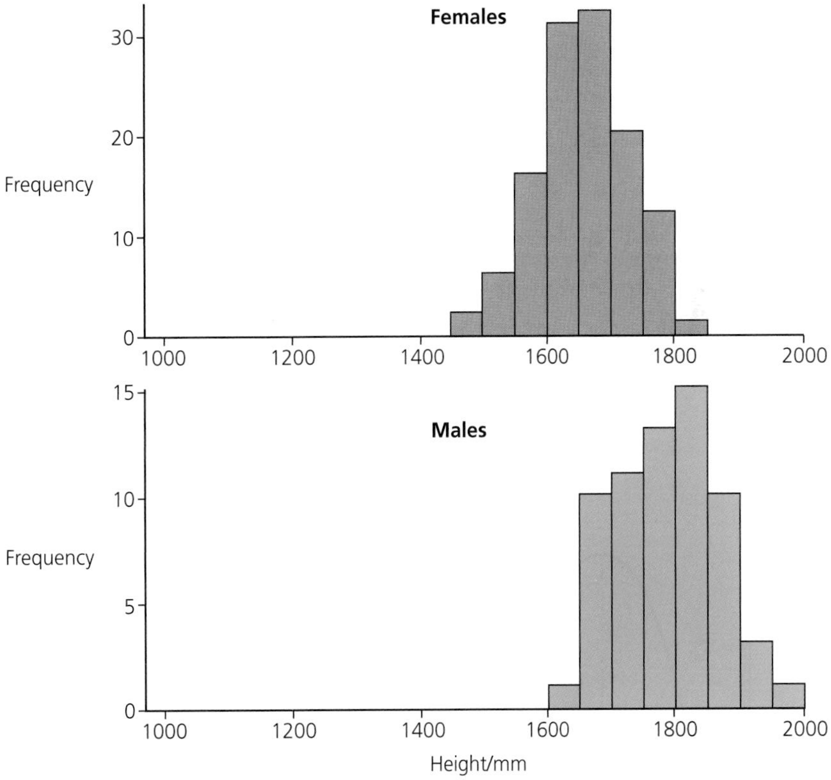

> **TIP**
> Genetic variation can be inherited or passed on to offspring. Environmental variation is not passed on.

**Figure 11.13** Frequency distributions showing the height of female and male students in a group of A-level biology students.

> **TEST YOURSELF**
> **10** Bacteria have been found in samples of ice that was formed in the Antarctic thousands of years ago. Some of these bacteria were found to be resistant to antibiotics. Suggest why.
> **11** Explain why cheetahs have a low genetic diversity and why this concerns scientists.
> **12** Seed banks contain seeds of the ancestors of modern crop plants. Suggest why these seeds are kept.

# Investigating variation

The organisms in a population seldom look exactly the same. Although they look broadly similar, they vary slightly in appearance, size and behaviour. Examples of variation include the stripe patterns among zebra, root length in carrots and feeding rate in fiddler crabs. Some kinds of variation can be measured and this is the sort of variation you will probably investigate.

Figure 11.13 may suggest to you the problems that we might have in finding the average height of humans in the UK. We could not measure every human in the country. Instead we would have to measure a small group of the population, called a sample. But would this sample be representative of the country as a whole? For example, if we had more females than males in our sample, this would give us a lower average height.

Any sample might not be representative of the population from which it is taken. This might happen for two reasons.

- Chance: what we commonly call 'luck'. We can reduce the effect of chance by taking several samples and finding their average value, or one very large sample.
- Sampling bias: this happens when the investigator, knowingly or unknowingly, chooses which measurements to include in the sample. We can reduce the effect of sampling bias using a random sampling technique. Random sampling is a technique of selecting the individuals in a sample that removes the investigator's choice, and ensures that the measurements are representative of the whole population and not affected by bias.

Look back to Figure 11.13. Each bar in the histogram shows the number of individuals falling into a particular category. This histogram is called a frequency distribution.

Figure 11.14 shows another frequency distribution. The sample used to make Figure 11.14 was much bigger than that for Figure 11.13, and the distribution is shown as a smooth curve rather than as a histogram. The frequency distribution in Figure 11.14 is of a special type, called a normal distribution. It has several important features, as follows.

- The value at the peak of the curve is the **mode**, or most frequent value. In a normal distribution, the peak is also the **median**, or middle value, and the **mean** value.
- The curve of a normal distribution is symmetrical, with 50% of the values below the peak (to its left) and 50% above the peak (to its right).
- Ninety five per cent of the values are within two standard deviations of the mean.

We have introduced two new terms here: **mean, median, mode** and **standard deviation**. Let's examine them further.

## Mean

The mean value is the correct term for what is sometimes referred to as the 'average'. We find it by adding up all the measurements we made and then dividing this total by the number of measurements we made. We can represent this in words as:

$$\text{Mean} = \frac{\text{sum of all measurements}}{\text{number of measurements}}$$

**TIP**

See Chapter 14 for how to calculate mean and standard deviation.

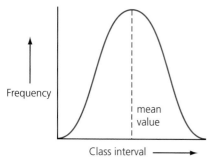

**Figure 11.14** A normal distribution curve.

In mathematical notation, this is written as:

$$\bar{x} = \frac{\Sigma x}{n}$$

where $\bar{x}$ (pronounced 'x bar') is the mean value, $\Sigma$ stands for 'sum of', $x$ represents each measurement made and $n$ is the number of measurements made.

## Standard deviation

The standard deviation is a measure of the spread of the data around the mean value. Think back to your practical work in class. You will have carried out an investigation to determine the effect of a variable, say temperature, on the activity of an enzyme-controlled reaction. When you did this, you probably used at least three replicate tubes at each temperature and calculated from them the mean time for the reaction to finish (the end time). Suppose you had found the end times in your three replicate tubes were 44, 45 and 46 seconds. You would probably have felt pleased with these results because they show little variation. You would have been confident in the **precision** of your result. Now suppose your end times were 25, 45 and 65 seconds. Although they give the same mean time as the first example, i.e. 45 seconds, you would probably not have been pleased with these results. The second set of results shows too much variation; they are less precise. You would have doubted that the mean value represented the true value you had tried to measure.

From these two examples, you will realise that the mean itself does not give us enough information about our sample. We want to know how spread out the results were that gave us our mean. If they are not spread out, we are more confident about their reliability than if they are very spread out. This is where the standard deviation is helpful, since it is a measure of the variation around the mean. Figure 11.15 shows two normal distribution curves; they represent the measurements made on two different samples. You can see that the two curves have the same mean but that they have different shapes. Curve a has a narrow spread of measurements, i.e. the measurements in this sample were very similar. Curve b has a much broader spread of measurements, i.e. there was much more variation in the measurements in this sample. Because there is greater variation within the sample in curve b, it will have a larger standard deviation than curve a.

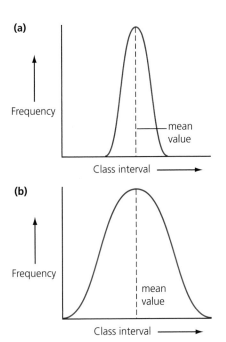

**Figure 11.15** These two normal distribution curves have the same mean value. Curve a, with the smaller spread of values, has a smaller standard deviation than curve b.

**TEST YOURSELF**

**13** Suggest one environmental factor that might account for the variation in the height of females in Figure 11.13.

**14** Why is it important that samples used to investigate variation are random samples?

**15** A student measured the heights of two different groups of people. Group A had a mean height of 185 cm and a standard deviation of 3.7 cm. Group B had a mean of 185 cm and a standard deviation of 7.5 cm. What does this information tell us about the two samples?

# EXAMPLE

Dog whelks are marine snails that live on rocky shores. Scientists measured the shell length of a random sample of adult dog whelks from a sheltered rocky shore and a rocky shore exposed to large waves. The shell lengths are given in Table 11.5.

1 Calculate the mean length for each sample of dog whelks
*The mean length for the sheltered shore sample is:*
*(22 + 24 + 35 + 30 + 32 + 21 + 26 + 31 + 37 + 24 + 34 + 28 + 29 + 36 + 34 + 28 + 22 + 34 + 23)/19 = 550/19 = 28.947 mm*
*which you should round to 28.9 mm.*
*The mean length for the exposed shore sample is*
*(19 + 22 + 23 + 24 + 21 + 16 + 19 + 21 + 22 + 16 + 23 + 15 + 25 + 21 + 24 + 21 + 23 + 24 + 18)/19 = 397/19 = 20.895 mm*
*Which you should round to 20.9 mm.*

2 Is there a difference in mean length for the dog whelks on the sheltered and exposed shores?
*Yes, the mean shell length on the sheltered shore is 8 mm greater than the mean shell length on the exposed shore.*

But simply finding the mean values does not necessarily tell you the whole story about the data in each sample. If the standard deviation is calculated for each set of measurements, a summary of the results can be given, as in Table 11.6.

3 What does the standard deviation tell you about the spread of the shell lengths in each sample?
*The standard deviations indicate that there is more variation on the measurements in the sheltered shore sample. If you imagine the data plotted as frequency histograms like Figure 11.13, the bars for the sheltered shore would have a larger spread, whereas the bars for the exposed shore would be more clustered together. The overall shapes of the bars would be similar to the two curves in Figure 11.14. This gives you more information than just looking at the mean values. The dog whelks from the exposed shore are smaller, but they also show much less variation in size.*

4 Suggest why there are these differences in the variation of shell length on the two shores
*One suggestion might be that greater wave action on the exposed shore means that there is a more critical optimum size for dog whelks. Dog whelks that are too small may lack sufficient ability to grip the rocks and if they are too large they may be more prone to being knocked off by the waves. Extreme sizes are less likely to survive in these conditions. On a sheltered shore shell size is less important and a wider range of sizes are able to survive. In addition, in sheltered conditions they may be able to spend more time feeding and so are able to grow larger.*

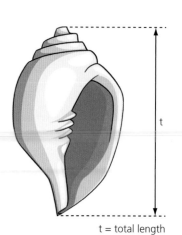

t = total length

**Figure 11.16** Dog whelk shell showing length measurements.

**Table 11.5** Shell lengths for dog whelk shells from a sheltered and an exposed rocky shore.

| Shell length on sheltered shore /mm | Shell length on exposed shore /mm |
|---|---|
| 22 | 19 |
| 24 | 22 |
| 35 | 23 |
| 30 | 24 |
| 32 | 21 |
| 21 | 16 |
| 26 | 19 |
| 31 | 21 |
| 37 | 22 |
| 24 | 16 |
| 34 | 23 |
| 28 | 15 |
| 29 | 25 |
| 36 | 21 |
| 34 | 24 |
| 28 | 21 |
| 22 | 23 |
| 34 | 24 |
| 23 | 18 |

**Table 11.6**

| | Mean shell length/mm | Standard deviation |
|---|---|---|
| Sheltered shore | 28.9 | ±5.25 |
| Exposed shore | 20.9 | ±2.98 |

# Practice questions

**1 a)** Complete the table to give the name of the term to which each
definition applies. (4)

| | |
|---|---|
| | Base that pairs with thymine in the DNA molecule |
| | The fixed position on a strand of DNA where a particular gene is found |
| | Protein associated with DNA in a eukaryotic chromosome |
| | A sequence of nucleotides within a gene that does not code for a polypeptide |

**b)** The figure shows a pair of homologous chromosomes from one
individual, as seen during prophase at the beginning of meiosis.

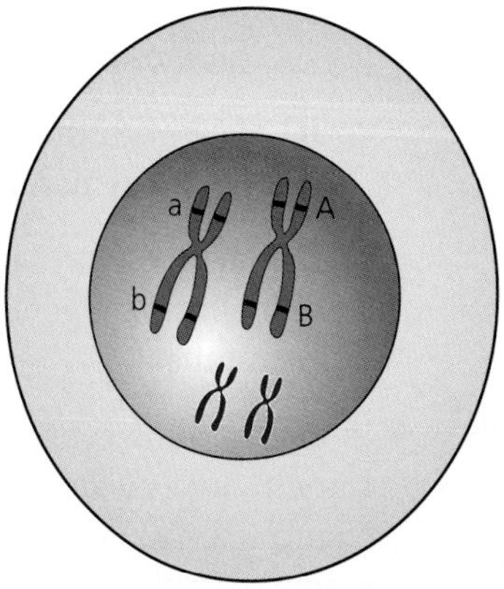

**i)** Give **two** pieces of evidence from the diagram that these
are homologous chromosomes. (2)

**ii)** What is the evidence that these chromosomes were drawn
during prophase at the start of meiosis? (2)

**c)** Most of the gametes from this individual contained alleles
A and B, or alleles a and b. However, a small number of gametes
contained alleles A and b, or a and B. Explain why. (2)

**2** A long-term study of antibiotic resistance was carried out between
1978 and 1993 on children suffering from middle ear infections. The
number of children given doses of antibiotic and the percentage of
resistant strains of bacteria identified were recorded each year. The data
are shown in the table.

| Year | Annual antibiotic usage/doses per 1000 people day$^{-1}$ | Resistant strains of bacteria/% |
|---|---|---|
| 1978 | 0.84 | 0 |
| 1979 | 0.92 | 2 |
| 1980 | 1.04 | 29 |
| 1981 | 0.98 | 46 |
| 1982 | 1.02 | 45 |
| 1983 | 1.03 | 58 |
| 1984 | 0.95 | 61 |
| 1985 | 1.12 | 60 |
| 1986 | 1.06 | 49 |
| 1987 | 1.14 | 59 |
| 1988 | 1.21 | 58 |
| 1989 | 1.28 | 71 |
| 1990 | 1.32 | 84 |
| 1991 | 1.31 | 79 |
| 1992 | 1.27 | 78 |
| 1993 | 1.28 | 91 |

**a)** The scientists concluded that increased antibiotic usage leads
to increased resistance in bacteria. Do these data support this
conclusion? Give reasons for your answer. (5)

**b)** Explain how natural selection results in antibiotic resistance in bacteria. *(4)*

**3 a) i)** What is a gamete? *(1)*

**ii)** How is a gamete different from a normal body cell? *(2)*

The diagram shows the life cycle of a fern. The numbers in the boxes represent the number of chromosomes in one cell at each stage.

**b) i)** Complete the empty boxes to show the chromosome number at each stage. *(2)*

**ii)** Mark with the letter M on the diagram a point at which meiosis occurs. *(1)*

**c)** Explain two ways in which cells produced by meiosis are genetically different from each other. *(4)*

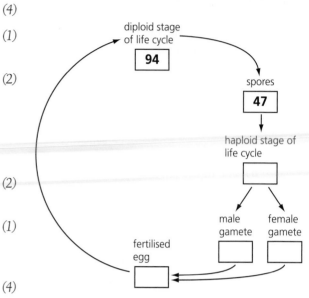

**4** In the late nineteenth century, Francis Galton, a half-cousin of Charles Darwin, investigated whether intelligence is inherited or affected by environmental factors. He created a family tree. Part of his family tree is shown in the figure.

**Chart showing the inheritance of intelligence**

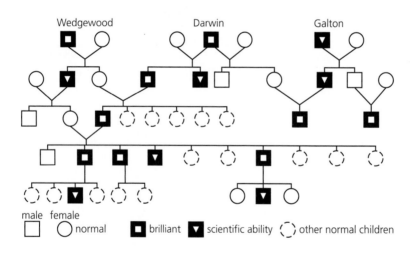

male female
□ ◯ normal   ■ brilliant   ▼ scientific ability   ⬚ other normal children

**a)** Galton concluded from this evidence that intelligence is inherited. Evaluate this conclusion. *(5)*

**b)** In a modern study, scientists wanted to find out whether height in humans is inherited. They measured the heights of dizygotic same-sex (non-identical) and monozygotic (identical) twins. All the twins grew up in the same home as their twin. The graphs show their results. Do these data show that height in humans is genetic or environmental? Give reasons for your answer. *(4)*

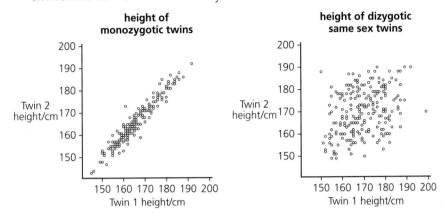

**5 a)** The Afrikaner population of Dutch settlers in South Africa is descended mainly from a few colonists. Scientists studying the present-day Afrikaner population have found that this population has an unusually high frequency of the gene that causes Huntington's disease. Suggest how this has occurred. *(3)*

**b)** The apple maggot fly originally fed on hawthorn fruits in the United States. About a hundred years ago, it started to become a serious pest of apple orchards and it is now a serious pest throughout the northern United States. Use your knowledge of natural selection to explain how this has happened. *(3)*

## Stretch and challenge

**6** The banded snails you saw in this chapter show a wide range of banding patterns on their shells as well as pink and yellow shells. Some have no bands at all. This means a large variety of different combinations of shell colour and banding pattern can be found in the population as a whole. This is known as a genetic polymorphism. Explain what the polymorphism is in *Cepaea nemoralis* and explain the cause of it.

# 12 Species and taxonomy

## TEST YOURSELF ON PRIOR KNOWLEDGE

1 Name two of the kingdoms that living organisms can be categorised into.
2 What is a DNA nucleotide made up of?
3 Give the complementary base pairs in DNA.
4 What is the maximum number of amino acids that a piece of DNA that is 420 base pairs long could code for?
5 What is functional RNA?

## Introduction

Other than both being plants, if you look at a London plane tree and a sacred lotus (see Figure 12.1) you would be forgiven for supposing that they are not very closely related.

One is a large tree with shiny leaves and flaky bark and the other is an aquatic plant with colourful flowers and pepperpot-like seed pods. They could not appear more different. Traditional classifications agreed, placing the lotus with water lilies in the family Nyphaeaceae and the plane tree in the family Platanaceae. But external appearances can be deceptive, especially among living organisms.

**Figure 12.1** (a) A London plane tree. (b) A sacred lotus.

More recent work has compared the base sequences of several key plant genes for hundreds of plant species and some of the results have been a real surprise. Based on an analysis of DNA rather than on outward appearance, roses are closely related to figs and nettles, orchids are not related to lilies but to yellow star grasses and, most surprising of all, the lotus is not related to the water lilies, which it most closely resembles, but to the plane trees that line many of London's streets.

Modern approaches to classification now routinely use molecular evidence alongside the appearance and behaviour of organisms, and a great deal of this research goes on in laboratories rather than in the storerooms and herbaria of museums. Reference collections in museums still play an important role in classification, as a tour behind the scenes at the Natural History Museum in London will confirm, but much of the work of a taxonomist now revolves around DNA and protein sequencing.

## Why do we classify organisms?

We like to organise and classify things. For example, a cookery book can have recipes grouped into starters and main courses, and meat and vegetarian dishes, and the weekly television guide divides programmes according to the day of the week and the channel. If knowledge wasn't classified in this way, we would never know where to look for information. Imagine, for example, a library full of books that were placed randomly on the shelves. If you wanted information on a particular topic, such as growing roses or the French Revolution, you would waste a lot of time trying to find it.

We really don't know how many different species of organisms there are, but it certainly runs into tens of millions. We need a system that classifies all the organisms we know about to try to establish evolutionary links. To be of real use the system must be universal – it has to be usable by biologists anywhere in the world.

# What is a species?

The system we use is based on dividing living organisms into species. To use this system we need to have a clear idea of exactly what we mean by a species. There are several things that we consider in defining a species.

First we need to look for similarities and differences. They might involve physical features. For example, in the blackbird and song thrush we see some similarities but differences in the colour of their feathers and their size.

These similarities and differences are reflected in their **binomials** (scientific names). Each species of organism has a unique binomial made up of two words. The binomial of the blackbird is *Turdus merula*. The blackbird is one of a larger grouping of species, the thrushes, which are all related to each other. The thrush group, or **genus** (plural **genera**), is called *Turdus* and all the different species of thrush found in Britain are named *Turdus something*. The song thrush is *Turdus philomelos* and another, the mistle thrush, is *Turdus viscivorus*. The blackbird is closely related to these two species of thrush so it has the same generic name, but its **species** name is *merula*.

## TIPS

- When you are writing a binomial, it is either written in italics or underlined. The generic or first part of the name starts with an upper-case letter, and the species or second part of the name is all lower case, e.g. *Homo sapiens*.
- Two or more species that share the same generic name are more closely related to one another than they are to species that have different generic names.

However, it is not just physical features that we should be looking at. The features that distinguish different species are controlled by genes, so we should expect to find differences in the DNA of different species or similarities in the DNA of the same species. Genes code for proteins and functional RNA (see Chapter 4, page 62), and so there will be differences in sequences of proteins such as the haemoglobins that transport oxygen in different species. There will also be differences in their RNA.

Another way that species differ is the diploid number of chromosomes in their cells. A rabbit has 44 chromosomes and a hare has 48, whereas a horse has 64 and a donkey has 62. If haploid gametes from two different species undergo fertilisation, the cell that forms has a different number of chromosomes to either parent. A horse gamete has 32 chromosomes whereas a donkey gamete has 31, so the cells of a mule, which is a **hybrid** of the two, have 63. Cells with an odd number of chromosomes are not viable and usually can't carry out successful meiosis because an odd number of chromosomes cannot form homologous pairs. This means that mules cannot produce gametes, so they are sterile.

This provides us with one way of defining a species. Two organisms belong to the same species if they are able to produce fertile offspring. The offspring of horses and donkeys are sterile, so they are two different species.

# When is a species not a species?

We can describe a species, then, as a group of organisms that share certain observable characteristics, and are able to produce fertile offspring. If we use this definition it ought to be easy to decide whether or not two organisms are separate species. Unfortunately, though, it is not always easy to decide whether or not two organisms are the *same* species.

## A thorny problem

Hawthorn is a common woody plant. There are two species of hawthorn: *Crataegus monogyna* (common hawthorn) and *Crataegus oxyacanthoides* (midland hawthorn). There are some obvious differences between these two species. Table 12.1 shows some of these.

**Table 12.1** Some differences between the two British species of hawthorn.

| Feature | *Crataegus monogyna* (common hawthorn) | *Crataegus oxyacanthoides* (midland hawthorn) |
|---|---|---|
| General appearance | | |
| Number of seeds in berry | One | Two |
| Shape of leaf | Many indentations | Few indentations |
| Hairs on leaf veins | Tufts of hairs present | No tufts of hairs present |
| Habitat | Along edges of woods and in open areas | In mature woods |

The evidence in the table suggests that *C. monogyna* and *C. oxyacanthoides* are different species. But are they? Hawthorns have always been planted as hedges. To grow the earliest hedges, farmers would have taken young hawthorn plants from nearby woodland and planted these. It is likely that they would have collected plants of both types.

Now look at the graph in Figure 12.2. It shows data about the leaves of hawthorn trees growing in a very old hedge, thought to have been planted over 900 years ago. The x-axis is an index of indentation of the leaves. This index was calculated by comparing the total depth of indentations with the length of the leaves. The more indentations there are on a leaf, the greater the value of the index of indentation. The y-axis shows the number of trees with each value.

We need to make sure that we understand the underlying biology before we look at the graph in detail.

- This hedge was planted over 900 years ago. How old are the hawthorn trees in the hedge now?

The answer to this is that we don't really know. It is very likely that all of the original hawthorn trees have died. Those present now are their descendants and they will have a range of ages. Some will be very old trees; some may be young.

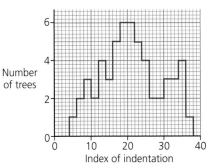

**Figure 12.2** Graph to show variation in leaf shape in the hawthorn trees growing in a hedge planted over 900 years ago.

The scientists who collected and analysed the data in the graph calculated the index of indentation by comparing the total depth of indentations with the length of the leaves.

● What was the advantage of presenting the data in this way?

The leaves on a particular tree vary in size. Size is bound to affect the total depth of leaf indentation: the bigger the leaf, the greater the depth of the indentations. Calculating an index like this enabled the scientists to compare leaves of different size.

The scientists calculated the index of indentation for a sample of *C. oxyacanthoides* trees growing in their natural environment. The mean value was 10.

● Explain why some of the trees growing in the hedge had values less than 10.

There will be variation in leaf indentation. When we say that the mean value for *C. oxyacanthoides* was 10, some trees will have a value less than 10 and some will have a value more than 10.

The scientists also calculated the index of indentation for a sample of *C. monogyna* trees growing in their natural environment. The mean value was 34.

● Explain why there are a large number of hawthorn trees in the hedge with an index between 10 and 34.

We can explain some of these intermediate values as being due to variation. A lot of these plants, however, are hybrids between *C. oxyacanthoides* and *C. monogyna*.

When we look at these hybrid plants in more detail, we find that they are intermediate in a wider range of characteristics. What is more, they are fertile and in turn produce fertile offspring.

● Are *C. oxyacanthoides* and *C. monogyna* different species?

This is a difficult question to answer. Table 12.1 shows us that there are some very obvious differences between the two types of hawthorn. But, when they are planted together, they can breed and produce fertile offspring. Using our definition they are the same species but, because they clearly look so different and hybrids are rare, they are currently regarded as separate species. Examples such as the two types of hawthorn, and the two types of duck shown in Figure 12.3, illustrate the problems that biologists have in defining species. Biologists also find that new evidence constantly leads them to reconsider how closely organisms are related. Sometimes this means that they have to change the binomials of organisms. You will read about an example of this, the bug orchid, on page 217.

Figure 12.3a shows a ruddy duck, *Oxyura jamaicensis*. Ruddy ducks were brought to the UK from North America. Then, in 1953, some of them escaped from captivity. Now they are widespread in western Europe. The white-headed duck, *Oxyura leucocephala*, seen in Figure 12.3b, is a native of Europe. It is also now a globally threatened species. Its numbers worldwide have fallen in the last hundred years from perhaps 100 000 birds to just over 5000.

**Figure 12.3** (a) The ruddy duck. (b) The white-headed duck.

White-headed ducks face extinction because they produce hybrids with ruddy ducks. Before this was known, the two ducks were thought to belong to different species. There are now about 550 white-headed ducks in the whole of Spain, the largest population in western Europe. To protect them, Spain decided to exterminate all their ruddy ducks. There are no white-headed ducks in the UK, and the policy is to kill all the ruddy ducks here to protect the Spanish white-headed ducks. But should we be doing this? This is the type of question that we have to ask ourselves as biologists.

# Sorting out species

In the system we use to classify different organisms, we put them into groups. These groups are based on things that organisms have in common and how they evolved. If two organisms have common features and if there is evidence that they also have the same ancestor, we assume that they are related, and put them into the same group. Often, but not always, the features they have in common reflect their evolutionary history.

Look at the organisms in Figure 12.4. We could divide them into those organisms that can fly and those that are unable to fly. This classification gives us no information about their evolutionary history. However, if we look carefully at the structure of their wings, we can see that, although an albatross wing and a penguin flipper are used for different purposes, they are really very similar. We can recognise the pattern of bones in an albatross wing as being the same as the pattern of bones found in a penguin flipper. This similarity points to the albatross and the penguin having a common ancestor at some stage in the distant past, and so being quite closely related.

**Figure 12.4** (a) A dragonfly.

(b) An albatross. It has very long wings and can fly huge distances.

(c) A penguin. It cannot fly, but swims using its flippers as paddles.

**Phylogenetic classification system** A system of classification based on evolutionary origins and relationships.

**Hierarchy** The placing of smaller groups within larger groups with no overlap between them.

A dragonfly's wing has a very different structure. Clearly, a dragonfly is not closely related to either an albatross or a penguin. A phylogenetic classification system makes use of features like wing bones that show a common evolutionary history. The classification system groups species together based on their evolutionary history. Similar species are grouped into genera, similar genera into families, and so on. A system like this, where smaller groups are placed into larger groups with no overlap between them, is called a hierarchy. The series of groups are called **taxa**. The highest ranked taxon is the domain, and biologists currently recognise three domains. One possible hierarchy is summarised in Figure 12.5 (overleaf), which shows the classification of the rabbit.

215

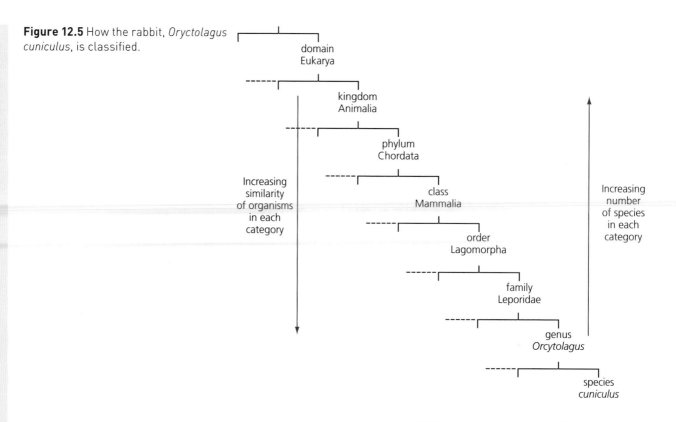

**Figure 12.5** How the rabbit, *Oryctolagus cuniculus*, is classified.

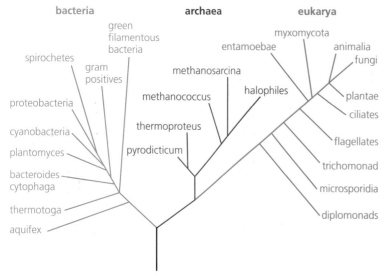

**Figure 12.6** Phylogenetic tree showing the three domains based on molecular similarities.

All living organisms are placed in one of three **domains**, the Bacteria, the Archaea and the Eukarya. These reflect some fundamental differences in the RNA sequences found in the organisms within these taxa. A phylogenetic tree represents the evolutionary relationships between taxa. Figure 12.6 shows a phylogenetic tree for the three domains based on molecular similarities suggesting that the Archaea are more closely related to the Eukarya than the Bacteria.

> **TIP**
> When you are writing the name of a group, the taxon should have a lower-case letter and the group name should have an upper-case letter; for example, domain Eukarya or kingdom Fungi.

## DNA, proteins and classification

The examples you have read about so far in this chapter show that biologists are not always certain whether organisms belong to the same or different species. Relying on physical features to sort organisms into different species can be misleading.

Biologists also have difficulties deciding how closely different species are related. For example, a wild orchid called the bug orchid originally had the scientific name *Orchis coriophora*, and that is the name you will find in many books. In natural conditions it forms hybrids, not just with closely related species, but with orchids of a different genus. Because of this, biologists now think they might have classified it wrongly. They have suggested that it ought to be placed in the same genus as the orchids with which it forms hybrids, so they have changed its scientific name to *Anacamptis coriophora*.

Do two species belong to the same genus? Are two families of organisms closely related? Biological research that finds the answers to questions such as these helps us to understand evolutionary relationships.

## TEST YOURSELF

**1** The scientific names of three wild cats are shown in the table. What do their scientific names tell you about how these wild cats are related to each other?

| Common name | Scientific name |
|---|---|
| Lion | *Panthera leo* |
| Leopard | *Panthera pardus* |
| Clouded leopard | *Neofelis nebulosa* |

**2** A horse and donkey can interbreed. The offspring is called a mule. Cells in a mule can undergo mitosis but not meiosis. Are a horse and a donkey different species? Give an explanation for your answer.

**3** Explain what a hierarchy is.

**4** What is a phylogenetic classification based on and what does it show?

The physical features that help us to classify an organism are determined mainly by its genes. A gene is a piece of DNA that codes for a protein, and proteins are the molecules that control the physical features of an organism.

In Chapter 4 you learned about DNA and its structure. Molecular biologists use machines to analyse DNA. They have worked out the complete DNA base sequences of a number of different organisms including humans. They have also found the DNA base sequences of particular genes in many different organisms.

In Chapter 5 you learned how DNA is copied in the process of replication. Errors sometimes arise when base sequences are being copied. When a base sequence in a gene has a copying error, we say there has been a **mutation**. A base may be added to the sequence, replaced by another base, or may be deleted altogether. If this happens in a body cell, it occurs only in that individual. If it happens to sex cells, the next generation inherits the change. Such mutations may either make no difference to the characteristics we see in a species, or they can cause the species to change very slowly over a period of many thousands of years. Either way, the DNA that codes for a particular protein in an organism alive today is slightly different from the DNA that coded for the same protein in its distant ancestor.

## Comparing DNA and mRNA base sequences

We can use computer software to compare DNA or mRNA base sequences in different organisms. If the sequences are very similar, it suggests that the organisms concerned are closely related and that they originated from a common ancestor relatively recently. If there are more differences between the sequences, it suggests that the organisms are not so closely related and probably originated from a common ancestor a longer time ago. Because mRNA is derived from the base sequence on DNA, sequencing mRNA gives the base sequence of DNA, and mRNA is sometimes easier than DNA to isolate from the cells of organisms.

### EXAMPLE

### Classifying whales

There are two groups of whales. Large whales such as fin whales and humpback whales do not have teeth. Instead, they have baleen plates, which are like huge combs that they use to filter small organisms from the water. They are put into one group. Dolphins and porpoises have teeth and are put into a second group. The sperm whale is a large whale, but it does not have the baleen plates. Instead, it has teeth, so, in the past, biologists classified sperm whales with the dolphins and porpoises.

Figure 12.7 shows this information in a diagram. The fin whale and humpback whale are closely related to each other and they split off from a common ancestor relatively recently. The bottlenose dolphin and the harbour porpoise are related in a similar way. The relationships shown in this diagram have been built up by looking at physical characteristics such as teeth. What do we find when we look at their DNA?

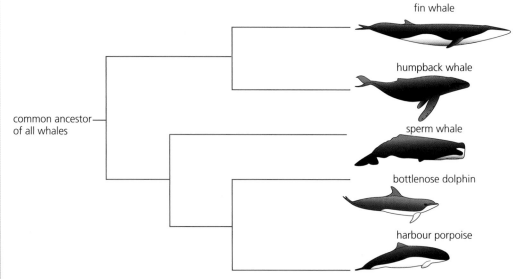

**Figure 12.7** This diagram shows how five species of whale could be related to each other.

Now look at Figure 12.8. It shows DNA base sequences in part of a gene that these different species of whales and dolphins have in common.

In this diagram, the bases are lined up with each other so that they match as closely as possible. Biologists use a computer to give the best possible match.

| fin whale | TAAACCCCAATAGTCA–CAAAACAAGACTATTCGCCAGAGTACTACTAGCAAC |
|---|---|
| humpback whale | TAAACCCTAATAGTCA–CAAAACAAGACTATTCGCCAGAGTACTACTAGCAAC |
| sperm whale | TAAACCCAGGTAGTCA–TAAAACAAGACTATTCGCCAGAGTACTACTAGCAAC |
| bottlenose dolphin | TAAACTTAAATAATCC–CAAAACAAGATTATTCGCCAGAGTACTATCGGCAAC |
| harbour porpoise | TAAACCTAAATAGTCC–TAAAACAAGACTATTCGCCAGAGTACTATCGGCAAC |

**Figure 12.8** Matching DNA base sequences in different whales and dolphins.

We can analyse the differences between these pieces of DNA by using a simple scoring system. We will start by comparing the humpback whale DNA with the DNA from the sperm whale. Some bases match. These have been highlighted in orange. We will score 1 for each match; these two sequences score 48. In other words, there are 48 matches between them. This tells us very little. To get a more detailed picture, we need to count up the matches between all the other pairs of species as well. These scores are shown in Table 12.2.

**Table 12.2** Similarities between DNA sequences in some whales and dolphins.

|  | Fin whale | Humpback whale | Sperm whale | Bottlenose dolphin | Harbour porpoise |
|---|---|---|---|---|---|
| Fin whale |  |  |  |  |  |
| Humpback whale | 51 |  |  |  |  |
| Sperm whale | 48 | 48 |  |  |  |
| Bottlenose dolphin | 43 | 43 | 41 |  |  |
| Harbour porpoise | 40 | 45 | 45 | 48 |  |

Let us work through this table and see what it tells us.

1  Using only the evidence from the table, which two species appear to be most closely related?
*The fin whale and the humpback whale are most closely related. The score of 51 tells us that they have 51 matching bases. The greater the numbers of matching bases the more closely are two species related.*

Here are the scientific names of some whales:

Fin whale               *Balaenoptera physalus*
Sperm whale            *Physeter catodon*
Southern right whale   *Eubalaena australis*
Northern right whale   *Eubalaena glacialis*

2  Suppose you analysed the same piece of DNA in these four species. Between which two would you expect the highest score, and why?
*The two right whales belong to the same genus, so they should be the most closely related to each other, and should have the highest score.*

The sperm whale is a large whale that has teeth. Figure 12.7 shows the sperm whale classified with the bottlenose dolphin and the harbour porpoise.

3  Does the evidence in the table suggest that this is the best way of classifying the sperm whale?
*If you look at the table carefully, you will see that the sperm whale has 48 matches with the fin whale and 48 matches with the humpback whale. It has fewer matches with the dolphin (41) and the porpoise (45). This seems to suggest that Figure 12.7 does not show the best way of classifying the sperm whale. It would be better to classify it in the way shown in Figure 12.9. This diagram suggests that the sperm whale is more closely related to the other large whales than it is to the bottlenose dolphin and the harbour porpoise.*

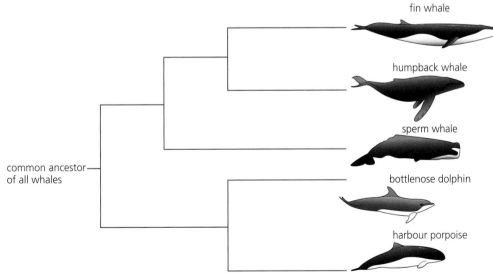

**Figure 12.9** An alternative way of showing how the whales and dolphins shown in Figure 12.7 may be classified.

**4** We have two different ways of ways of classifying the sperm whale. Which is the correct one?

*We have to be very careful about how we interpret evidence about DNA base sequences. Here we have looked only at the similarities and differences in one piece of DNA. Sometimes the evidence we get from* *another piece of DNA may suggest something else. In coming to their conclusions, biologists have to weigh up data from different sources. In this case, all the available evidence now suggests that Figure 12.9 is a better interpretation than Figure 12.7 of how the five whales should be classified.*

## ACTIVITY

### Another way of looking at differences in DNA: comparing mRNA sequence

Tree shrews (Figure 12.10) are mammals from Southeast Asia. Because of their small size and short reproductive cycle they have been suggested as a possible alternative to non-human primates in biomedical research. For example, they have been used to investigate the way in which the influenza H5N1, or bird flu virus, infects organisms. Scientists hope that this will provide useful information about how the H5N1 virus infects humans. But their suitability depends on how closely they are related to primates. There has been some disagreement among scientists as to whether tree shrews are insectivores or primates because they share features of both groups.

Scientists have investigated this by comparing the base sequences of some of their mRNA molecules to those of other primates. One study investigated tree shrew mRNA for interleukin 7 (IL-7), a protein involved in the immune system. First the scientists found the sequence for tree shrew IL-7 mRNA from an online database. Then they used the database to find the mRNA sequence for the same protein for another 14 mammal species and used computer software to compare all the sequences. The results are shown in Figure 12.11.

**1** Does Figure 12.11 suggest that the tree shrew is a good alternative to non-human primates in biomedical research?
**2** The mouse and the pig are both currently used as alternatives to non-human primates in biomedical research. Use Figure 12.11 to explain why.

**Figure 12.10** Tree shrews share the characteristics of insectivores and primates.

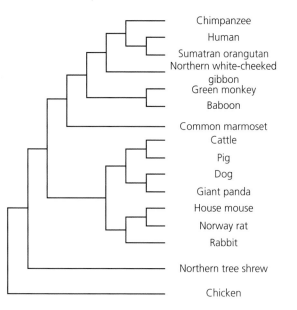

**Figure 12.11** The relationship of tree shrews to some other animals in terms of their interleukin 7 mRNA base sequences.

# Comparing amino acid sequences

The human body contains many different proteins. You will already have come across some of them, such as enzymes, carrier molecules and antibodies. Each protein is made up from a particular sequence of amino acids. The amino acid sequence reflects the base sequence on the DNA, but proteins are sometimes easier to isolate from the cells of organisms than DNA. If we look at a protein present in several species, we may see that the sequence of amino acids in this protein differs slightly from one species to another. We can find out more about how closely humans are related to different species of apes by comparing the amino acids in their proteins.

We will start by looking at human haemoglobin. Every haemoglobin molecule in an adult human contains four polypeptides: two identical α chains and two identical β chains. Each α chain consists of 141 amino acids joined to each other by peptide bonds. The β chains are slightly longer. They each contain 146 amino acids. Together we have 141 + 146, or 287 amino acid positions that we can compare. Surprisingly, if we take the four species that we looked at in the previous section, 282 of these amino acids are in exactly the same sequences in all four. There are differences in the positions of only five amino acids, and these are shown in Table 12.3.

**Table 12.3** Differences in amino acids in the haemoglobin of humans and three species of ape.

| Species | α chain | | | β chain | |
| --- | --- | --- | --- | --- | --- |
| | Position 11 | Position 23 | Position 87 | Position 104 | Position 115 |
| Human | Alanine | Glutamic acid | Threonine | Arginine | Proline |
| Chimpanzee | Alanine | Glutamic acid | Threonine | Arginine | Proline |
| Gorilla | Alanine | Asparagine | Threonine | Lysine | Proline |
| Orang utan | Threonine | Asparagine | Lysine | Arginine | Glutamine |

In the table, the shading shows the places where amino acids are the same as on the human α and β chains. You can see that both chains in chimpanzee haemoglobin have exactly the same sequences of amino acids as both chains in human haemoglobin.

We need to be careful how we interpret this. It doesn't mean that humans and chimpanzees are the same species! Remember, the data are for only one molecule, haemoglobin. A much more likely explanation is that it takes a very long time for differences in the amino acid sequence of haemoglobin to evolve. It is possible that only a few million years have passed since humans and chimpanzees split apart from a common ancestor. Perhaps this is too short a time for differences in their haemoglobin to have evolved.

## Finding relationships with RNA sequence data

Figure 12.12 shows the phylogenetic relationship between five species of fruit fly from the genus *Drosophila*.

1 How many millions of years ago did *Drosophila yakuba* and *Drosophila melanogaster* last share a common ancestor?
*They last had a common ancestor 8 million years ago, because this is the time of the last point at which they were connected on the tree.*

2 Which two fruit fly species are most closely related?
*Drosophila mauritania and Drosophila sechellia, because they share a common branch of the tree.*

3 This diagram is based on mRNA sequence data rather than DNA sequence data. Explain why it is often easier to obtain mRNA sequence data from tissue samples.
*mRNA molecules can be found in the cytoplasm of cells, and there are often many molecules of the same mRNA present because of repeated transcription. Isolating these is often easier than finding a gene in the DNA in the nucleus.*

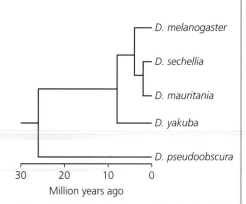

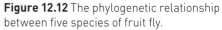

**Figure 12.12** The phylogenetic relationship between five species of fruit fly.

## TEST YOURSELF

5 Proteins are polymers of amino acids. Give two ways in which the primary structure of proteins may be different.

6 Looking at differences in the amino acid sequence of haemoglobin is not useful for studying classification of all organisms. Explain why.

7 Which three types of sequence data can be used to investigate the relationships between species?

# Practice questions

**1 a)** Complete the table to show the classification of the dandelion, *Taraxacum officinale*. (2)

|  | Eukarya |
|---|---|
| kingdom | Plants |
|  | Anthophyta |
|  | Magnoliopsida |
|  | Asterales |
|  | Compositae |
| genus |  |
| species |  |

**b)** Some scientists found a group of dandelions growing on the side of a hill. These dandelions had much smaller leaves than the dandelions they found in other areas. They wondered if this group of dandelions was a new species. How could they find out whether these dandelions were a different species from those with larger leaves? (2)

**2** The table shows the number of amino acid differences in the protein cytochrome *c*, taken from different organisms. Cytochrome *c* is a protein used in aerobic respiration.

| Organism | Number of amino acid differences |
|---|---|
| Human | 0 |
| Monkey | 1 |
| Pig | 10 |
| Dog | 11 |
| Duck | 11 |
| Turtle | 15 |
| Tuna | 21 |
| Moth | 31 |
| Yeast | 51 |

**a) i)** Give one reason why cytochrome *c* might be a better protein to use for examining evolutionary relationships than haemoglobin. (1)

**ii)** These data suggest that monkeys are more closely related to humans than pigs. Explain how. (2)

**b)** A student looked at these data and suggested that they show that dogs are closely related to ducks. Is this a valid conclusion? Explain your answer. (2)

**3** Scientists wanted to investigate the evolutionary relationships between several different species of viper, *Bitis* spp. This was done as follows.

**A** They extracted pure samples of the protein albumen from the blood of each species.

**B** They injected a sample of the albumen into rabbits and extracted antibodies against each type of albumen.

**C** They mixed samples of each of the antibodies produced with each of the different kinds of albumen.

**D** They used these results to produce the diagram shown in the figure.

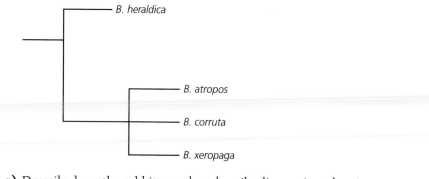

**a)** Describe how the rabbits produced antibodies against the viper albumin when it was injected into their blood in stage B. *(4)*

**b) i)** Describe what the results show about the relationships between *Bitis corruta, Bitis atropos* and *Bitis heraldica*. *(2)*

   **ii)** Describe how the results from stage C could be used to produce the results shown in the diagram. *(3)*

**c) i)** At least two rabbits were used to produce each kind of antibody in stage B. Suggest why. *(1)*

   **ii)** If the scientists compared the base sequences of the gene that codes for albumen from these different species of snake they might have obtained slightly different results. Suggest why. *(1)*

### Stretch and challenge

**4** The original work that underpins the idea of the three domains was carried out by a scientist called Carl Woese. What made him come to the conclusion that the kingdom Prokaryotae should be separated into two domains?

# Biodiversity within a community

**TEST YOURSELF ON PRIOR KNOWLEDGE**

1 What is a community?
2 Give an example of an animal or plant species that is adapted to a particular environment.
3 Describe how two plant species might compete with one another.
4 Explain why animals are dependent on plants.

## Introduction

The photograph on the left shows a wood at the edge of Hauxley Nature Reserve in Northumberland. The area the wood occupies was once part of an open-cast coal mine. When the mine closed, the whole coal-mining area was landscaped to include a lake with islands. Then in 1983 the Northumberland Wildlife Trust bought the land to develop into a wildlife reserve. The Trust decided to plant trees and create the small wood that you see in the photograph.

Questions the Trust asked were:

- What trees should we plant?
- Where should we plant them?

These decisions would influence the number and kinds of other organisms that came to live in the wood, including birds, insects and mammals, and the Trust wanted to make the mix of species in their wood – its biodiversity – as wide as possible.

**Habitat** The environment in which an organism or population of organisms usually lives.

Wherever habitats such as at Hauxley Nature Reserve's wood are developed, compromises need to be made. One compromise is that making a wood will inevitably destroy open ground, which is the habitat of species that do not live in woods. Also, particular species of tree have specific ecological needs and this limits the choice of suitable trees. Alder and willow grow in moist soil; birch will not grow in heavy shade; and beech requires alkaline soils. Clearly, not all species of trees will grow in a particular area. In the Trust's project and others like it, the second compromise is to plant different tree species to encourage a wide range of animal species, yet to plant only tree species that will grow well in the area.

# Investigating biodiversity

Scientists carry out investigations that can help to answer questions such as those posed for Hauxley Wood. We will look at the results of some of the research that has been done on woodland birds. These results are shown in Figure 13.1. We have a much better picture of the factors determining the different species of birds that live in a wood than we have for most other groups of animals and plants. This is because many scientists have studied bird populations. Also, it is generally true that a wood that supports many different species of birds will also support many different species of other organisms. This makes studies on birds very useful when selecting types of trees and deciding where to plant them to form a new wood.

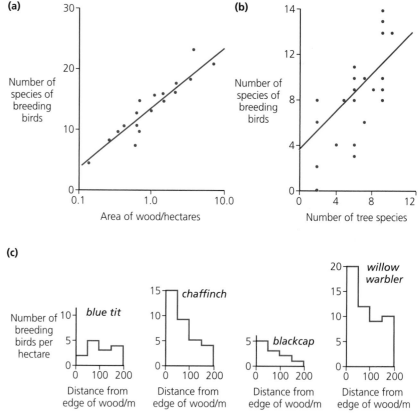

**Figure 13.1** Graphs showing the results of some research on woodland birds. (a) Number of species found in woods of different sizes. (b) The relationship between the number of species of birds and the number of species of tree growing in a wood. (c) The relationship between distance from the middle to the edge of the wood on the breeding populations of blue tit, chaffinch, blackcap and willow warbler.

We will start by looking at the data in Figure 13.1a. The graph shows the number of species of birds found in woods of different sizes. On the *x*-axis in this graph, the numbers for the size of the wood have been plotted on a logarithmic scale. This means that the values on the *x*-axis increase by a factor of 10 each time. Instead of going up in equal increments (0, 0.2, 0.4, 0.6 hectares, etc.) the scale goes from 0.1 to 1.0 to 10.0 hectares. The advantage in plotting data this way is that we can show a far greater range of sizes and see more easily if there is a correlation.

A lot of studies have been carried out on the effect that wood size has on the number of bird species. They have all produced similar results to those in Figure 13.1a. The line drawn through the points is called the line of best fit. It shows the overall trend of the data. Since the slope is upwards, the trend is upwards, so there is a **positive correlation** (see page 249) between number of species and wood area. This means that we can conclude that as the size of the wood increases so does the number of bird species.

Figure 13.1b shows data collected from a second investigation. In this graph, the number of breeding species of birds has been plotted against the number of different species of tree present. Again, there is a clear correlation between the two. The more species of trees that are present, the more species of birds that breed.

Finally, we look at Figure 13.1c. The scientists who collected the data for these graphs investigated the effect of distance from the edge of the wood on the number of times that particular species of birds were recorded. The graphs show the results for four different species. Notice that records for blue tits are spread roughly evenly. For the other three species, however, more birds are found near the edge of the wood, and there are fewer in the interior.

The results from scientific investigations such as those shown in the graphs help us to make decisions about conserving biodiversity, such as those made by the Northumberland Wildlife Trust at Hauxley Nature Reserve. The results suggest three things, as follows.

- As large an area as possible should be planted with trees because there is a clear correlation between the area of a wood and the number of species of birds.
- Different tree species should be planted. Again, there is a clear correlation between the number of species of tree and the number of breeding bird species.
- The wood should have an irregular shape or should have open areas within it. This will allow for more woodland edge, and this is the habitat favoured by most species of bird.

# Measuring diversity

Biodiversity can be considered at a range of scales, from a small habitat such as a ditch or patch of meadow to entire countries and even the Earth. Look up the word **diversity** in a dictionary and you will see that it simply means 'being different'. So species diversity means 'the mix of different species'. The simplest way of measuring species diversity in a community is to count the number of species present. This is called species richness. A community with more species has a richer mixture of species. The

**Community** A group of interacting populations of different species living in the same place at the same time.

**Species richness** The number of different species in a community.

data in both Figures 13.1a and 13.1b shows species richness for the bird community in a wood. Producing a list of species and then simply counting them up can, however, be misleading if you are comparing communities.

If you were to look walk through a typical deciduous wood in the UK and look carefully at the trees you would find quite a lot of species present. Some will be more common than others but they would all count to the same extent in a species richness score. Look at Table 13.1 (opposite). The species richness is 8 but it would still be 8 even if there were 100 oak trees rather than just 6. The numbers of each species give an idea of the evenness of the mix of the community. A wood dominated by oak trees would not have such an even mix as the one in Table 13.1.

**Index of diversity** An index giving the relationship between the number of species in a community and the number of individuals in each species.

A more useful measurement of species diversity for the trees in the wood would take into account both the number of species present (richness) and the number of individuals of each species (evenness). We call such a measurement an index of diversity. It describes the relationship between the number of species present and how each species contributes to the total number of organisms in that community. Because of the fairly even mix of quite a few tree species, this index would be high for the wood in question. Let us look at how a biologist who gathered information about the diversity of trees in the wood used the formula for an index of diversity.

## Calculating an index of diversity

Here is the formula for calculating an index of diversity, $d$:

$$d = \frac{N(N-1)}{\Sigma n(n-1)}$$

> **TIP**
> As you will see in other formulae in Chapter 14, the symbol $\Sigma$ means 'the sum of'. So in this formula it means the sum of all the different values of $n \times (n-1)$.

In this formula:

$N$ = the total number of organisms in the community

$n$ = the number of organisms of each species in the community.

> **TIP**
> If you look up the term index of diversity elsewhere you will find that there are many kinds. The formula above is a type of Simpson's index of diversity. There are even two versions of this, each of which can be expressed in three different ways. You need to stick to the formula shown here. With this formula, the larger the number you obtain the greater the diversity.

The biologist walked along a path through the wood and recorded all trees within 5 m of the path along a 50 m length. His results are shown in Table 13.1.

**Table 13.1** The numbers of trees of different species along a path through a wood.

| Species | Number of trees |
|---------|----------------|
| Beech | 2 |
| Cherry | 1 |
| Hawthorn | 1 |
| Hazel | 4 |
| Holly | 7 |
| Lime | 14 |
| Oak | 6 |
| Rowan | 1 |
| Total | 36 |

Then he calculated the index of diversity ($d$). We can substitute the figures for $N$ and $n$ from the data in the table:

$$d = \frac{36(36-1)}{2(2-1) + 1(1-1) + 1(1-1) + 4(4-1) + 7(7-1) + 14(14-1) + 6(6-1) + 1(1-1)}$$

$$d = \frac{36 \times 35}{(2 \times 1) + (1 \times 0) + (1 \times 0) + (4 \times 3) + (7 \times 6) + (14 \times 13) + (6 \times 5) + (1 \times 0)}$$

$$d = \frac{1260}{268} = 4.7$$

### How would the biologist use this information?

On its own, the index of diversity of the wood does not mean a lot. Its value is that it allows us to compare the diversity of the trees in this wood with diversity of trees in different woods and other habitats. Biologists concerned with species diversity find it very valuable to make comparisons like this.

A higher index of diversity indicates a richer community where ideal conditions allow more species to be more equally successful. A lower index of diversity indicates fewer successful species and may indicate more challenging or stressful conditions such as a restricted range of food sources, fewer habitats or pollution. In such conditions, one or two well-adapted species may thrive, but they will dominate the community. If an index of diversity decreases over time, that is especially significant.

## EXAMPLE

### Investigating heathland diversity

Lowland heath is a very diverse community of plants and animals and a unique UK habitat. The dominant plant species are heather, gorse and ling and the plant community supports a variety of insects such as the heath grasshopper and the potter wasp. Bracken is found on many heaths and is a potentially invasive species that is shade tolerant.

A scientist investigated the diversity of plants in an area of heath. The table shows the data the scientist collected.

| Species | Number of plants/m² |
|---------|---------------------|
| Ling | 3 |
| Bell heather | 8 |
| Heath bedstraw | 6 |
| Bird's foot trefoil | 2 |
| Bracken | 5 |
| Tormentil | 5 |
| Mat-grass | 10 |

1 Calculate an index of diversity using the formula:

$$d = \frac{N(N-1)}{\Sigma n(n-1)}$$

where:

$N$ = total number of organisms of all species in the community

$n$ = total number of organisms of a particular species in the community.

*To find N, total up the number of each species in the table. This comes to 39. To find $\Sigma n(n-1)$, take each species and multiply the number found by the number found minus 1. Remember that $\Sigma$ means add each of these together. So $\Sigma n(n-1) = 3(3-1) + 8(8-1) + 6(6-1) + 2(2-1) + 5(5-1) + 5(5-1) + 10(10-1)$. This comes to 222. So:*

*d = 39(39 − 1)/222*

*d = 1482/222*

*d = 6.68*

*This is a relatively high index of diversity.*

2 If heathland is not grazed by animals, trees such as birch can become established. Suggest why this would reduce the diversity of insects.

*Trees are taller than the low-growing plant species in the table. They will create shade, which would reduce the ability of many of the plant species to compete for light so they would not grow effectively. Bracken can tolerate shade more than the other species so it would spread and become dominant, further out-competing the other plant species for light. The variety of different food sources and habitats for insects found in the diverse plant community would both be reduced. In turn, this would reduce the number of species of insect able to survive and reproduce on the heath so the diversity of insects would also fall.*

**TEST YOURSELF**

1 Suggest two reasons why a large wood often contains many more bird species than a small wood.

2 A student was asked to compare the diversity of birds in the middle of a wood and at the edge of the same wood. How should she do this?

3 Willow trees are planted in some areas to provide biofuel. These trees are planted close together. Would you expect the diversity of trees in such a willow plantation to be higher or lower than that for the wood discussed earlier? Explain how you arrived at your answer.

4 The index of diversity for the invertebrate community in a stream declines at the point of a treated sewage outfall and then slowly recovers further downstream. Explain why.

# Farming and biodiversity

Since the Second World War there have been huge changes in the way that land has been farmed. Agricultural machines have become larger and more powerful, and to work efficiently they need very big fields. Farmers rear more productive varieties of livestock, and grow more productive varieties of plants. They control insect pests and weeds with chemical pesticides. As a result of these changes, farming has become more intensive, with more food produced per hectare. Also, farms have become more specialised, concentrating for example on just crops or just livestock. All this adds up to larger quantities of food being produced more cheaply.

Farmers are under pressure to use all available land for food production. In the drier east of England, for example, most farmers now concentrate on growing crops such as cereals. They have removed hedges to create larger fields, drained wetter grazing land and have filled in most farm ponds because they no longer have farm animals that need water. Changes in farming practices like these have reduced the diversity of wild plants and animals found on farmland. We will look at some specific examples of how changes in farming have affected diversity.

## Improving grazing land

Intensive production of milk requires good quality grass for grazing. Ideal grazing pasture to support large numbers milk cows has to be fast growing and nutritious. A dairy cow can eat 19 kg of dry plant biomass per day.

For high milk yield, the quality of the food is just as important as the quantity. If there is too much **cellulose** in the food, it takes longer to digest (see Chapter 1, page 7) and so the cow eats less. It therefore has less energy available to put into milk production. Figure 13.2 shows that high-quality food (feed quality index of 1.0 or less) can generate as much as 25 kg of milk per cow per day.

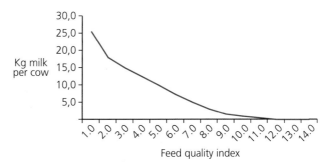
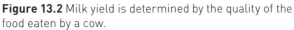

**Figure 13.2** Milk yield is determined by the quality of the food eaten by a cow.

The best-quality grazing for milk can be obtained by growing a mixture of ryegrass and white clover. These species will only grow on well-drained and fertilised land. Where possible, dairy farmers improve traditional pasture by draining and re-seeding. Whereas a few other plant species might colonise this grassland over time, the community has a low species richness. Regular use of fertiliser encourages new, vigorous growth of these two species at the expense of any others so although the food quality is high, the diversity of plants remains low (Figure 13.3).

**Figure 13.3** Improved grassland, showing predominantly ryegrass and white clover.

Unimproved grazing pasture, especially where the land is poorly drained, looks very different (Figure 13.4). It contains a mixture of more than one grass species and a number of other flowering plant species, many adapted specifically to tolerate damp conditions. Such wet grassland may support 10–15 plant species in a square metre.

**Figure 13.4** Unimproved wet grassland; notice how many different types of plants there are.

These different species and their different heights, flowers, seeds and leaves offer a far wider range of food sources and habitats to insects such as grasshoppers, butterflies and beetles. In turn, these insects provide food for insectivorous birds and mammals. The diversity of insects, mammals and birds on and around unimproved wet grassland is higher. But dairy cows grazed on such land would be eating older, tougher grass and a variety of other stalky plants with a higher cellulose content which would not allow them to produce as much milk.

Because of grassland improvement for dairy farming, large areas of grazing land have been improved to support high milk yields. There is very little wet grassland left in some parts of the country and the specialised plant species adapted to their damp conditions have become uncommon.

Estimates suggest that 97% of flower-rich grassland has been lost since 1930. It is thought that this could be one factor behind the current decline in animal diversity of intensively farmed countryside. Any remaining pockets of wet grassland now form **biodiversity hotspots** in the agricultural landscape (Figure 13.5).

**Figure 13.5** A pocket of unimproved grassland in an agricultural landscape. Notice where the biodiversity hotspot is.

## Spring sowing of cereal crops

For bird species that live on arable land, the two most important events in the farming year are ploughing and harvesting. When a field is ploughed, the soil is turned over and many invertebrates are brought to the surface. Large numbers of birds are often attracted to newly ploughed fields to feed on these invertebrates.

This supply of food, however, only lasts for a day or two after ploughing. Harvesting is also a time when food is plentiful. Seed-eating species of birds can feed on spilt grain and seeds from weeds growing in the crop. Insect-feeding species also benefit because soil-living invertebrates become more accessible once the cover provided by the crop plants has been removed.

One of the main changes in arable farming is a switch from spring planting to autumn planting of cereal crops. Look at Table 13.2 in the Activity box below. You will see that spring sowing involves ploughing the soil in early March. The grain is sown in March or early April. It germinates in April and is ready for harvest in September. Fields are left with stubble over winter before being ploughed again the following March. Autumn sowing involves ploughing immediately after a harvest in July. The grain is then sown in September and starts to germinate. The young plants grow rapidly once spring comes, and the grain is ready for harvesting in late June or early July. Some bird species have declined as a result of this change. This may lead to a reduction in the diversity of farmland birds in the future.

**Arable land** Land used to grow crops.

**TIP**
Improving grassland and sowing crops in the autumn instead of spring are simply examples to illustrate the impacts of changing farming techniques on biodiversity, which you might be asked to interpret in an exam. **You do not need to memorise these examples.**

**ACTIVITY**

### The effect of autumn sowing of cereal on rooks

Rooks are omnivorous birds; their food includes plant and animal material: they often feed in fields, eating soil invertebrates; they also eat seeds, including newly planted and spilt grain. Surveys of rook numbers were carried out by biologists and showed a decrease, particularly in eastern England. The biologists thought that this decrease was due to the change from spring to autumn sowing of grain crops.

1 Look at the first column in Table 13.2. This shows information about spring sowing. In which months would you be unlikely to find rooks feeding in arable fields?
2 When they were feeding in spring-sown fields, on what do you think the rooks would be feeding in (a) October and (b) March?
3 With autumn-sown crops, for what proportion of the year is food available in arable fields?
4 Between 1975 and 1980, rook numbers decreased in south-east England but remained more or less the same elsewhere. Use the data in the table to suggest an explanation for this.

**Table 13.2** This table summarises the differences in the timing of the spring and autumn sowing of cereal. The shading shows when rooks rely heavily on arable fields for their food.

| Spring sowing | Month | Autumn sowing |
|---|---|---|
| Stubble | January | |
| Stubble | February | |
| Ploughing | March | |
| Sowing and germination | April | |
| Sowing and germination | May | |
| | June | |
| | July | Harvesting |
| | August | Ploughing |
| Harvesting | September | Sowing and germination |
| | October | Sowing and germination |
| Stubble | November | |
| Stubble | December | |

233

## Organic farming

Organic farms, where pesticide and inorganic fertiliser use is restricted, are thought to benefit biodiversity. Apart from the fact that the food they produce is free from harmful chemicals, one of the reasons that customers may pay a premium for organic produce is the idea that they are supporting farming practices that support higher plant and animal diversity on and around organic farms. Although this is a widely held view, there is surprisingly little rigorous scientific evidence to support it.

### ACTIVITY

#### The effect of organic farming on butterflies

One study in Sweden compared the butterflies found on organic farms with those found on conventional farms. Twelve pairs of farms were selected for the study. Six pairs were in a mixed landscape where the mean percentage of arable land was only 15%. The other six pairs were located in a more uniform arable landscape where the mean percentage of land used for growing crops was 70%. In each case, an organic farm was paired with a conventional farm of similar area. The scientists tried to match pairs of farms as closely as possible in terms of landscape feature such as hedges, stone walls and patches of woodland.

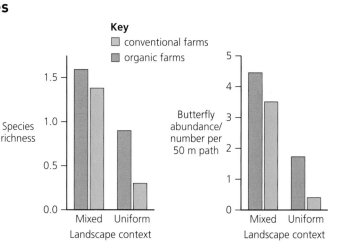

**Figure 13.6** Butterfly abundance and species richness in relation to farming practice.

**1** Why were the pairs of farms matched as closely as possible for landscape features?

**2** What was the reason for selecting six pairs of farms in a mixed landscape and six pairs in a more uniform landscape?

The results for the butterfly surveys are shown in Figure 13.6.

**3** What effect do organic farming practices have on (a) butterfly species richness and (b) butterfly abundance in mixed compared to uniform landscapes?

**4** Why might organic practices have had more impact on butterfly species richness in uniform landscapes?

**5** Suggest how the restrictions on pesticide and inorganic fertiliser use on organic farms might explain the differences in butterfly species richness and abundance.

**6** The scientists did not calculate an index of diversity in their study. Could you use information from the graphs to calculate an index of diversity for each group of farms? Explain your answer.

# Disappearing bumblebees

Recent concern has been expressed over the possible link between intensive crop production and the extinction or near extinction of some bumblebee species. Some 24% of the 68 species of bumblebee found in Europe are threatened with extinction. This includes two species found in the UK, the great yellow bumblebee (*Bombus distinguendus*) and the shrill carder bee (*Bombus sylvarum*). A third UK species, the short-haired bumblebee, was declared extinct in 2000, although a recent attempt at re-introduction has been made in Kent.

This rapid reduction in the diversity of bumblebees is causing concern. It is not just because scientists are interested in bumblebees. Bumblebees are vital pollinators of arable crops such as oilseed rape, peas, strawberries and apples. Insect pollinators are estimated to contribute over £400 million to the UK economy every year. Bumblebees play a large part in this free ecosystem service. Without it the cost of some fruit and vegetables would rise considerably.

Ideal bumblebee habitats have a high diversity of plants. This ensures a supply of pollen and nectar at different and overlapping times throughout the bumblebees' feeding season. In agricultural areas, ideal habitats include species-rich grassland, field margins, hedgerows and ditches. Farmers have to manage these habitats sensitively to help bumblebee species survive (Figure 13.7).

**Figure 13.7** This species-rich field margin is suitable for bumblebees.

Bumblebees depend on a high diversity of flowering plants for a reliable source of food, whereas wild plant species depend on bumblebees for their pollination. A variety of plant species is threatened by the potential loss of bumblebee species, including arable crops. The biodiversity of bumblebees and plants are interlinked.

A laboratory study highlighted another possible factor in the decline of bumblebees. This investigated the impact of some pesticides on bumblebee foraging effectiveness. It found that chronic (long-term) exposure of bumblebees to two neonicotinoid pesticides impairs their ability to navigate and fly to find food.

Neonicotinoids are a relatively new kind of insecticide that affects the nervous system of insects, resulting in paralysis and death. They protect the young crop plants from insect pests from the very start of their life and are now widely used to treat seeds before planting. It is estimated that neonicotinoids have improved winter wheat yields in the UK by as much as 20%. Oilseed rape production now depends on them to combat pests such as the rape flea beetle. However, when used as seed treatments, neonicotinoids move to the pollen and nectar of the adult plants through **translocation** (see Chapter 10, pages 185–189), which is how they may affect bees and other pollinators as well as the target pest species.

The laboratory study gave rise to calls for these sorts of insecticides to be banned. This caused widespread concern among farmers. But there is no conclusive evidence from preliminary field trials to support the laboratory study and it is clear that the pesticides are extremely useful to farmers. This example clearly illustrates the difficulty of balancing the interests of productive arable farming with biodiversity conservation.

## TEST YOURSELF

**9** Give two reasons why organic farming is thought to increase biodiversity.

**10** Explain why maintaining the diversity of pollinators on arable farmland is so important.

**11** Why are neonicotinoid pesticides so useful to arable farmers?

**12** How might neonicotinoid pesticides used on seeds be causing problems for bumblebees?

## TIP

When you write about environmental issues such as those described in this chapter, remember that you are a scientist and you should write as a scientist. Use scientific terms and use them accurately. It is not a good idea to write, for example, about animals 'losing their homes' and 'having nowhere to live' because their habitat has been lost.

# Practice questions

**1** Scientists investigated the number of different bird species present in hedges of different lengths. The graph shows their results. Black circles represent hedges surrounding conventional fields, and white circles represent hedges surrounding organic fields where no pesticides had been used.

**a)** What is the relationship between the length of hedges and the number of bird species present in them? *(2)*

**b)** Suggest why the length of a hedge might have an effect on the number of bird species present. *(2)*

**c)** The scientists counted the number of different species present rather than calculating the index of diversity. Was this a good decision? Give reasons for your answer. *(3)*

**d)** Do these data show that organic farming is better for encouraging the diversity of birds? Give reasons for your answer. *(4)*

**2** Two students carried out a survey of bird species in two different woodlands in the same area of the country on the same day. They made a list of all the birds they saw or heard singing over a period of 1 hour in each woodland. Their results are shown in the table.

| Species | Number at site A | Number at site B |
|---|---|---|
| Blackbird | 7 | 3 |
| Blue tit | 2 | 5 |
| Bullfinch | 0 | 1 |
| Carrion crow | 1 | 0 |
| Chaffinch | 3 | 0 |
| Chiffchaff | 0 | 2 |
| Goldfinch | 2 | 0 |
| Greenfinch | 0 | 2 |
| Great tit | 4 | 2 |
| Magpie | 0 | 3 |
| Robin | 4 | 6 |
| Starling | 2 | 7 |
| Tree sparrow | 12 | 8 |
| Wood pigeon | 6 | 4 |

**a) i)** Give one additional variable that the students should have kept the same, explaining the reason for this. *(2)*

**ii)** Explain one possible limitation of this method and the effect this might have on the results obtained. *(2)*

**b) i)** Calculate an index of diversity for site A. *(2)*

**ii)** Suggest an advantage of calculating an index of diversity rather than just counting the number of birds seen or heard. *(1)*

**3 a)** What is meant by biodiversity? (2)

The figure shows the changes in the relative abundance of different groups of species in the UK from 1970 to 2010. The abundance of each group of species was measured relative to the first year they were sampled. The numbers in brackets indicate the number of species sampled in each group of organisms.

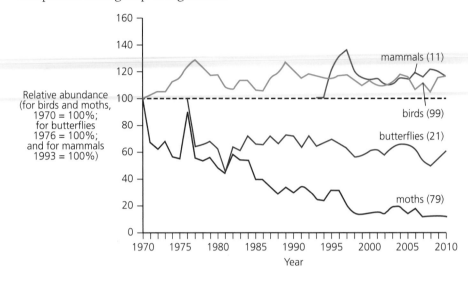

**b)** A student concluded that moth populations were much lower in 2010 than butterfly populations. Explain two reasons why this conclusion might be wrong. (4)

**c)** Explain how the relative abundance of mammals in 1998 would have been calculated. (3)

## Stretch and challenge

**4** Conservationists set up nature reserves such as Hauxley Wood to maintain biodiversity as high as possible. In order to achieve this, what can seem to be destructive activities are carried out, such as cutting down trees and bushes, putting grazing animals onto the reserve and cutting reeds in ponds. Explain how these activities can help to maintain biodiversity rather than reduce it.

# Developing mathematical skills

## Using units

In any measurements or calculations, using units correctly is critical. You need to be familiar with the common units used in biology and confident in converting between units, for example from millimetres (mm) to micrometres (μm) or from cubic centimetres ($cm^3$) to cubic millimetres ($mm^3$).

There are a thousand millimetres in a metre (m), a thousand micrometres in a millimetre and a thousand nanometres (nm) in a micrometre. There are also a thousand cubic millimetres in a cubic centimetre and a thousand cubic centimetres in a cubic decimetre ($dm^3$).

Since these are the units you are most likely to need to convert between, the 'rule of thousands' is a helpful one. But there are some important exceptions such as the centimetre. There are, of course, only 100 cm in a metre and 10 mm in a centimetre.

### TIP
Try to avoid using centimetres for measuring lengths. Use millimetres and then you are less likely to make a mistake if you need to convert them.

Sometimes units are combined in a rate such as how fast oxygen is being produced ($mm^3\,minute^{-1}$) or how fast an animal is moving ($cm\,second^{-1}$). It is easy to forget to combine the units when you give an answer following a rate calculation. Sixty is the important number to remember for converting units of time. There are 60 minutes in an hour and 60 seconds in a minute.

For some rates, the first unit is a count such as the number of breaths in a breathing rate or the number of beats in a heart rate. In this case, the correct units are $beats\,minute^{-1}$ or $breaths\,minute^{-1}$.

# Calculating areas, volumes and circumferences

You need to be confident with arithmetic calculations for finding the size and surface area of biological structures such as cells, organs and whole organisms. You also need to be able to give answers in the correct units and with the appropriate number of decimal places. It is also very useful to be able to recognise when your answer is outside of the expected range so that you realise when you may have made a mistake.

## Calculating areas of circles

Areas of rectangles are calculated by multiplying the lengths of the two sides whereas areas of circles are calculated by multiplying the square of the radius by $\pi$ (so area of a circle = $\pi r^2$). Area calculations are often used to find the surface area of an organism or a cell such as a red blood cell.

If you assume that a red blood cell has a flat surface rather than a concave one (Figure 14.1), the approximate surface area of a red blood cell can be calculated. In Chapter 9 you saw that a red blood cell has a diameter of $7\,\mu m$. The radius is half the diameter, which is $3.5\,\mu m$. One side of a typical red blood cell therefore has an area of $3.5^2 \times \pi = 38.48\,\mu m^2$. Note that the units are micrometres squared because this is an area measurement.

But this is the area of just one side of the red blood cell. Since it has two sides, the approximate total surface area of a typical red blood cell is $38.48 \times 2 = 76.96\,\mu m^2$.

### Decimal places and rounding

When you carry out calculations like finding the surface area of a red blood cell, the answer shown on your calculator will often include a number of **decimal places**. This is the number of figures after the decimal point. 76.96 has two decimal places. In this case, giving the answer to the nearest whole micrometre is quite sufficient, because the original diameter of the red blood cell was only given as $7\,\mu m$. So the answer should be **rounded** and it becomes $77\,\mu m^2$ rather than $76.96\,\mu m^2$.

You may also need to calculate areas of circles when considering the cross-**sectional areas** of blood vessels. Chapter 9 tells you that an average human capillary has a diameter of $8\,\mu m$. This means that its total cross-sectional area would be $4^2 \times \pi = 50.26\,\mu m^2$, which should be rounded to $50\,\mu m^2$. This calculation would be useful if you needed to compare the cross-sectional areas of different types of blood vessel.

Table 14.1 shows the mean diameters of the different types of blood vessel in a bat's wing. For each vessel, the cross-sectional area has been calculated by multiplying the square of the radius by $\pi$. You can see how much larger the cross-sectional area of a vein is compared to a capillary.

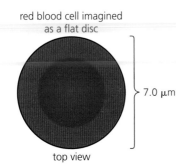

**Figure 14.1** A red blood cell can be imagined as a flat two-sided disc.

red blood cell imagined as a flat disc

$7.0\,\mu m$

top view

> **TIP**
> Rounding should be done using the rule that below half is rounded down and half or more is rounded up, so 1.24 becomes 1.2 and 1.25 becomes 1.3.

**Table 14.1** Mean diameters and calculated cross-sectional areas of the blood vessels in a bat wing.

| Vessel | Mean diameter/$\mu m$ | Cross-sectional area/$\mu m^2$ |
|---|---|---|
| Artery | 52.6 | 2170 |
| Arteriole | 19.0 | 284 |
| Capillary | 3.70 | 10.8 |
| Venule | 21.0 | 346 |
| Vein | 76.2 | 4560 |

### Using significant figures

The mean diameters have also all been shown to three **significant figures**. This means that in Table 14.1, 3.7 is shown as 3.70. This indicates that it is exactly 3.7 rather than a value rounded to one decimal place.

Significant figures in a number include all the non-zero digits, any zeros between non-zero digits and, in numbers containing a decimal point, all zeros written to the right of the digits. The number of significant figures in a measurement gives an indication of its uncertainty.

Because the mean diameters all have three significant figures, the calculated cross-sectional areas should only be shown to a maximum of three significant figures. This is why the cross-sectional area for the capillary is the only answer shown with a decimal place.

The cross-sectional area for the artery is $26.3 \times \pi = 2173\,\mu m^2$, but 2173 has four significant figures which would give the answer less uncertainty than the original measurement. This is why the answer is shown in Table 14.1 with three significant figures as 2170. The final 3 has been replaced with a zero, which is not a significant figure.

The cross-sectional area for the arteriole is $9.5 \times \pi = 283.5\,\mu m^2$ but 283.5 has four significant figures which would also give the answer less uncertainty than the original measurement. It is shown in Table 14.1 with three significant figures as 284. It has been rounded up from 283 because the last digit is 0.5, which is half or more.

## Calculating areas of cubes and spheres

Different sized cubes are often used to compare imaginary organisms of different sizes. The surface area of a cube is calculated by finding the area of one side and multiplying that by six (Figure 14.2), because it has six sides. So a cube of side length 8 mm has a surface area of $(8 \times 8) \times 6 = 384\,mm^2$. A cube of side length 16 mm has a surface area of $(16 \times 16) \times 6 = 1536\,mm^2$. Notice that although the side length has been doubled, the surface area has increased four times. You should always estimate roughly what the answer should be for a calculation like this so that you can spot if your answer is way out from what you are expecting and look for your mistake.

> **TIP**
>
> When you do calculations using measurements, you should limit the accuracy of your answers to that of the least accurate measurement.

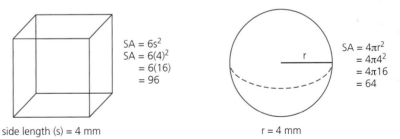

$$SA = 6s^2$$
$$SA = 6(4)^2$$
$$= 6(16)$$
$$= 96$$

$$SA = 4\pi r^2$$
$$= 4\pi 4^2$$
$$= 4\pi 16$$
$$= 64$$

side length (s) = 4 mm                  r = 4 mm

**Figure 14.2** Calculating the surface area of (a) a cube and (b) a sphere.

In the same way that cubes can represent different sized organisms, spheres can be used to represent cells of different sizes. You need to remember that:

Area of a sphere = $4 \times \pi \times radius^2$

Again, this means that if the radius of a cell is doubled, the surface area increases by a factor of four (Figure 14.2).

## Calculating volumes

Being able to calculate the volumes of cubes and spheres is also useful (Figure 14.3).

Volume of a cube = breadth × width × height

Volume of a sphere = × $\pi$ × $r^3$

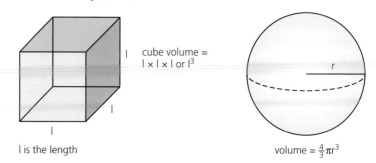

cube volume =
l × l × l or $l^3$

l is the length

volume = $\frac{4}{3}\pi r^3$

**Figure 14.3** Calculating the volume of (a) a cube and (b) a sphere.

Apart from calculating the volumes of cubes and spheres, you should also be able to calculate the volume of a cylinder. This is the cross-sectional area of the cylinder multiplied by its height.

Volume of a cylinder = ($\pi$ × $r^2$) × h

This is useful when using a potometer. The volume of water taken up by a leafy stem is measured by finding how far a bubble moves along a capillary tube. The water in the tube forms a cylinder shape (Figure 14.4), so the volume of water taken up is equal to the distance moved by the bubble (the length of the cylinder) multiplied by the cross-sectional area of the capillary tube **lumen**.

The cross-sectional area of the lumen is $\pi$ × $r^2$. So to find the volume of water taken up, you need to know the radius of the capillary tube (or its diameter so you can halve it). The radius of the capillary tube will most probably be in millimetres, and you should also measure the distance the bubble moves in millimetres so the answer will be in cubic millimetres. If you have timed how long the bubble took to move this distance, you could give the answer as a rate by dividing the volume by time in minutes. The appropriate units would then be $mm^3\,minute^{-1}$.

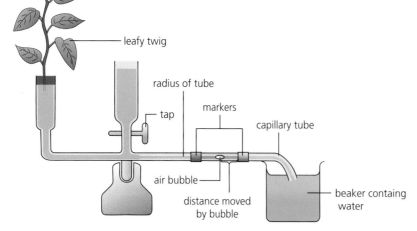

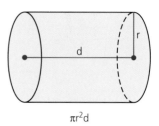

$\pi r^2 d$

**Figure 14.4** (a) Potometer showing the cylinder of water between the markers that has been taken up.

(b) The volume of water in the cylinder can be calculated by knowing the radius of the capillary tube.

## Calculating circumferences

The circumference of a circle is found by multiplying the diameter by $\pi$. If a red blood cell has a diameter of $7\,\mu m$, then its circumference is $7 \times \pi$, which is $21.99\,\mu m$, which should be rounded to $22\,\mu m$.

Knowing the formula for circumference is also useful when investigating the effect of transpiration rate on the diameter of tree trunks. In Chapter 10 you saw that when a tree is transpiring rapidly the tension in the xylem vessels pulls in the walls of the vessels just a little and causes the tree trunk to shrink very slightly.

It is difficult to measure the diameter of a tree trunk directly, but instruments called dendrographs can be used to measure the circumference. They have a band that is stretched tightly around the trunk, attached to a sensor. If the circumference increases or decreases, the sensor can measure the change. In rubber trees, mean daytime trunk shrinkage has been found to be $312\,\mu m$ near to the base of the trunk and as much as $569\,\mu m$ 2 m up the trunk.

If

$$\text{Circumference} = \text{diameter} \times \pi$$

Then

$$\text{Diameter} = \frac{\text{circumference}}{\pi}$$

This means that near the ground the rubber tree trunk diameter decreases during the day by as much as $\frac{312}{\pi} = 99.3\,\mu m$. Further up the trunk it decreases by $\frac{569}{\pi} = 181\,\mu m$. Note that because the mean measured changes in circumference were given to three significant figures, the calculated diameters have only been given to three significant figures.

# Using ratios and percentages

## Ratios

**Ratios** can be used to relate one attribute of an organism to another. For example, **surface area to volume ratios** are helpful in understanding why larger organisms need specialised exchange surfaces. Using the formulae from the previous section, a sphere of radius $10\,mm$ has a volume equal to $\frac{4}{3} \times \pi \times 10^3 = 4188.79\,mm^3$ and a surface area equal to $4 \times \pi \times 10^2 = 1256.64\,mm^2$

This means it has a surface area to volume ratio of $1257:4189$ which simplifies to $1:3.33$. The ratio is **simplified** by making the first number 1, so the second number is $\frac{1}{1257} \times 4189 = 3.33$. The ratio tells you that the numerical value of the volume is about three times that of the surface area. Imagine these numbers represent the surface area and volume of an imaginary cell. If the cell radius is doubled the volume now equals $\frac{4}{3} \times \pi \times 20^3 = 33510.32\,mm^3$ and the surface area now equals $4 \times \pi \times 20^2 = 5026.55\,mm^2$. So the surface area to volume ratio is now only $5026:33510$, which simplifies to $1:6.67$.

243

The numerical value of the surface area is now less than one sixth of that of the volume. This shows how, if cells or organisms are larger, the area of their surface does not keep up with the increase in their volume. This can mean that their surface is insufficient to supply enough materials and remove enough waste materials or heat, as seen in Chapter 7.

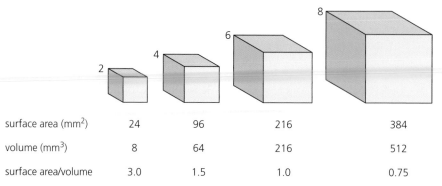

| surface area (mm²) | 24 | 96 | 216 | 384 |
| volume (mm³) | 8 | 64 | 216 | 512 |
| surface area/volume | 3.0 | 1.5 | 1.0 | 0.75 |

**Figure 14.5** Relationship between different sized cubes and their surface area to volume ratios.

### Standard form

When you divided 1 by 1257 to find the ratio in the previous example, the answer displayed on a calculator is $7.96 \times 10^{-04}$. This way of writing the number is called **standard form** and means that the decimal place is actually four positions to the left of where it is shown, in other words 0.000796, or **ordinary form**. This is a very small number.

Standard form for large numbers moves the decimal place in the opposite direction. 150000 becomes $1.5 \times 10^5$ in standard form; in other words, the decimal place is actually five places to the right.

Standard form avoids having to write out all the zeros in small or large numbers. However, you must be very careful to keep the same number of significant figures (see page 241) when converting numbers to and from standard form. A value of 0.0040 indicates that the value is exactly 0.004 rather than a rounded value. The last zero is a significant figure as well as the figure four so 0.0040 has two significant figures.

If you were converting 0.0040 to standard form, you would need to include the last zero by writing it as $4.0 \times 10^{-3}$. Both significant figures are shown in the standard form.

## Percentages

You will come across **percentages**, and especially the idea of a **percentage change**, quite often in biology.

Percentages should normally always add up to 100%. If you know the percentage of some of the bases in a length of DNA, you can calculate the percentage of their complementary bases. Table 14.2 shows an example where you have to calculate the missing values.

> **TIP**
> When using standard form, there can only be **one** digit to the left of the decimal point, so:
> 150000 must be written as $1.5 \times 10^5$, **not** as $15 \times 10^4$ or $150 \times 10^3$.

**Table 14.2** Percentages of bases in two complementary DNA strands.

|  | A | T | C | G |
|---|---|---|---|---|
| Strand 1 | 20 | 35 |  |  |
| Strand 2 |  |  | 30 |  |

Adenine (A) is complementary to thymine (T), so if there is 20% A in one strand, there must be 20% T in the other. If there is 35% T in one, there must be 35% A in the other. However, finding the values for cytosine (C) and guanine (G) is slightly harder. If there is 30% C on one strand there must be 30% G on the other. Both rows must add up to 100%, so the remaining values for G must both be 15%. Table 14.3 shows the answers in bold.

**Table 14.3** The missing percentages are shown in bold.

|  | A | T | C | G |
|---|---|---|---|---|
| Strand 1 | 20 | 35 | **15** | **30** |
| Strand 2 | **35** | **20** | 30 | **15** |

Expressing information as a percentage sometimes allows a valid comparison. For example, if you wanted to compare the incidence of heart disease in Scotland with that of England, you could not directly compare the number of people with heart disease, because the population of Scotland is smaller than the population of England. Instead, you would need to calculate the percentage of people with heart disease in each country. These percentages would then be directly comparable.

## Calculating percentage change

If you are collecting data in an experiment where the starting values for a measurement are all different, you can calculate percentage changes to make sure your results are comparable. When investigating the increase or decrease in mass of potato cylinders in different sucrose concentrations, the starting mass of all the cylinders will all be slightly different, even if you try and cut them the same size. So any increase or decrease in mass can be expressed as a percentage of the original mass, and all the results become directly comparable.

If a potato cylinder has an initial mass of 32 g which increased to 35 g in dilute sucrose solution, then the increase in mass is 35 − 32 = 3 g. The percentage increase is therefore $\frac{3}{32} \times 100 = 9.37\%$ which is correctly rounded up to 9.4%.

A quick estimate suggests that 3 is roughly a tenth of 32, so the percentage increase should be approximately 10%. If not, you may have made a mistake.

## Fractions

You may come across a situation involving **fractions** when looking at cells undergoing mitosis. The **mitotic index** is a way of expressing the amount of cell division taking place in a tissue. In a sample of cells, the fraction undergoing mitosis gives an indication of how much the tissue is actively dividing.

$$\text{Mitotic index} = \frac{\text{the number of cells in mitosis}}{\text{total number of cells counted}}$$

Cells in any stage of mitosis can be identified because their chromosomes are visible (Chapter 5). Other cells are counted as being in interphase. So if a sample of 200 cells contains eight in prophase, 14 in metaphase, five in anaphase and nine in telophase, the mitotic index is:

$$= \frac{8 + 14 + 5 + 9}{200} = \frac{36}{200}$$

This fraction can also be given as a decimal answer $\left(\frac{36}{200} = 0.18\right)$ or converted to a percentage by multiplying by 100 (0.18 × 100 = 18%).

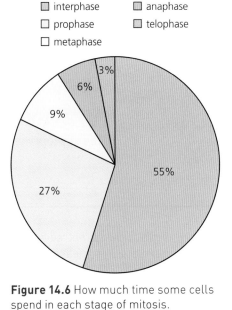

- interphase
- prophase
- metaphase
- anaphase
- telophase

3%
6%
9%
55%
27%

**Figure 14.6** How much time some cells spend in each stage of mitosis.

245

Figure 14.6 shows the percentage of time a sample of cells spent in each phase of mitosis. In this case, the mitotic index would be:

$$= \frac{27 + 9 + 6 + 3}{100} = \frac{45}{100}$$

The mitotic index can be used to find how long mitosis takes. If the whole cell cycle takes 22 hours and the mitotic index is 0.18, then mitosis takes $22 \times 60 \times 0.18$ minutes, which is 237.6 minutes.

### Percentage error

When you make measurement in practical work, the uncertainty in the measurement can be expressed as a percentage error. The percentage error in the measurements obtained from different pieces of apparatus can be calculated by dividing the uncertainty by the measured value and multiplying by 100.

A volume of 25 cm$^3$ measured with a measuring cylinder has an error of $\pm 0.5$ cm$^3$. The percentage error for this piece of apparatus is $\frac{0.5}{25} \times 100 = 2\%$. A mass of 0.120 g measured using a balance with an error of $\pm 0.001$ g. The percentage error is $\frac{0.001}{0.120} \times 100 = 0.833\%$.

## Dealing with orders of magnitude

Drawings and images of small structures such as cells and organelles are often larger than reality. Micrographs (photographs taken using a microscope) are magnified thousands of times. The magnification of the micrograph is usually given as a magnification factor such as ×10 000. This means that the image is 10 000 times larger than the object really is. You might be asked to calculate the actual size of something in the image, such as the actual width of a mitochondrion or the diameter of a cell. The method for doing this is always the same, regardless of the object or the magnification.

Measure the length of the structure in the image in millimetres. To convert it to micrometres, multiply by 1000. This is the image length. To find the real length of the structure, divide the image length you have found in micrometres by the magnification you have been given using the formula:

$$\text{Size of real object} = \frac{\text{size of image}}{\text{magnification}}$$

This will give you a real length in micrometres. Try this for the crista A–B in Figure 14.7. For microscopic structures, micrometres are the appropriate unit to use for the answer.

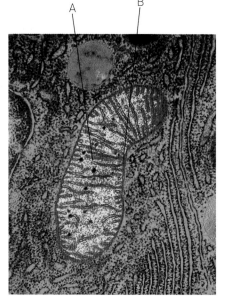

A          B

**Figure 14.7** Electron micrograph of mitochondrion (×24 000).

**TIP**

Always convert your measured length in millimetres to micrometres by multiplying by a thousand before doing anything else. Do not measure in centimetres. Always measure on the image in millimetres. This helps to avoid decimal place errors.

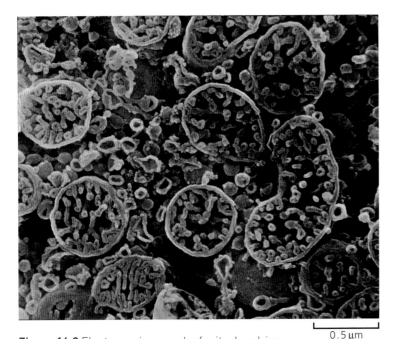

0.5 μm

**Figure 14.8** Electron micrograph of mitochondrion with scale bar.

To calculate the magnification of a structure, use the formula:

$$\text{Magnification} = \frac{\text{image length}}{\text{real length}}$$

You shouldn't need to remember this because you should realise that magnification is how big something looks (its image length) compared with how big it really is (its real length). The answer should be given in the form ×2000 rather than just the number.

If there is a **scale bar** on the micrograph, this is the real length (see Figure 14.8). Measure it with your ruler in millimetres and convert to micrometres. Then divide your measurement by the real length, the value on the scale bar. This gives the magnification factor.

# Using symbols and equations

You need to know the meaning of the mathematical symbols shown in Table 14.4 and when to use them.

You may also have to solve expressions such as $4^n$ where $n$ is a **power**. It might be to calculate how many combinations of three bases, or DNA triplets, are possible using four different bases. In this case, the expression is $4^3$ and the answer is $4 \times 4 \times 4 = 64$. The same sort of expression can be used to find the number of possible combinations of chromosomes following meiosis. If a fruit fly has 16 chromosomes and there is a homologous pair of each, then the number of possible combinations is $16^2$ or $16 \times 16$, which is 256.

There are a few equations that you need to be able to rearrange. Often it is only necessary to put the known values into the equation to calculate an answer, as in an index of diversity (see Chapter 13).

$$\text{Index of diversity, } d = \frac{N(N-1)}{\Sigma n(n-1)}$$

Where $\Sigma$ means 'the sum of', $N$ is the total number of organisms in the community and $n$ is the number of organisms of each species in the community. All that has to be done here is to use the data for the species in the community to put the right numbers into the equation.

Sometimes, it is necessary to rearrange the equation in order to find the answer. This was done with the magnification factor equation in the previous section. Magnification = image length/real length can be rearranged as

**Table 14.4** The meaning of some symbols used in equations.

| Symbol | Meaning |
|--------|---------|
| = | is equal to |
| < | is less than |
| << | is much less than |
| ≤ | is less than or equal to |
| > | is greater than |
| >> | is much greater than |
| ≥ | is more than or equal to |
| ∝ | is proportional to |
| ≈ | is roughly similar to |

**TIP**

You might be asked to use these equations in questions, but you do not need to learn them.

**TIP**

Although pulmonary ventilation rate is a rate, it is defined as the volume of air breathed in and out in a minute, so the correct units are just dm³. You do not need to say per minute.

**TIPS**

• If you are drawing your own scattergram, you should try to use a similar scale on each axis so that the results are spread out equally in both directions.

• You should always remember that just because two variables are correlated, it doesn't mean that one *causes* the other.

real length = image length/magnification depending on what you already know and what you need to find.

Two equations you will have seen in Chapter 7 and Chapter 9 are

Pulmonary ventilation rate (PVR) = breathing rate (BR) × tidal volume (TV)

Cardiac output (CO) = heart rate (HR) × stroke volume (SV)

These can be both rearranged depending on which values you know and which you need to find out. If you know cardiac output and heart rate, stroke volume = cardiac output/heart rate. But if you know cardiac output and stroke volume the equation can be rearranged so that heart rate = cardiac output/stroke volume.

Pulmonary ventilation rate is the volume breathed in and out in a minute. So if you are given pulmonary ventilation rate as 6 dm³ and tidal volume as 0.5 dm³, you can calculate the breathing rate from pulmonary ventilation rate/tidal volume = $\frac{6}{0.5}$ = 12 breaths per minute. But if you were given the breathing rate and pulmonary ventilation rate instead, the tidal volume is found from pulmonary ventilation rate/breathing rate = $\frac{6}{12}$ = 0.5 dm³.

## Plotting and using graphs

Drawing a graph helps you to see the relationship between two variables much more clearly than looking at data in a table. The two types of graph you will see most frequently are scattergrams and line graphs.

### Scattergrams

The graph in Figure 14.9 is a **scattergram**. You draw a scattergram when you want to know if there is a relationship between two variables. Figure 14.9 shows if there is a relationship between cigarette consumption and coronary heart disease. Smoking is a risk factor for coronary heart disease (see Chapter 9).

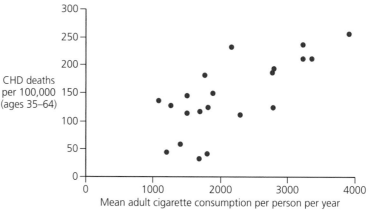

**Figure 14.9** Scattergram showing smoking as a risk factor for coronary heart disease.

The independent variable (cigarette consumption) is plotted on the *x*-axis, and the dependent variable (the incidence of coronary heart disease) is plotted on the *y*-axis.

When you draw a scattergram, ensure that you label the axes and include units. Someone else should be able to look at the scattergram and know exactly what it shows without any further explanation. Plot each point as a dot with circle round it or a cross. Draw (or imagine) a line of best fit. You can imagine the line of best fit for Figure 14.9.

### Correlation

If your line of best fit slopes upwards, you can say that there is a **positive correlation** between the two variables. In other words, as one variable increases, so does the other. If your line of best fit slopes downwards, you can say that there is **a negative correlation**. In this case, as one variable increases, the other decreases. Sometimes the line of best fit is horizontal, sometimes it is vertical and sometimes it is completely impossible to draw a line of best fit. In these cases, all you can say is that there is no correlation. In the case of Figure 14.9, you can say that there is a positive correlation between smoking and the risk of coronary heart disease.

## Line graphs

You will come across a lot of line graphs in biology. You use a line graph to show how one factor varies as you change another. You might draw line graphs for the results of enzyme rate experiments, for investigating the permeability of beetroot cell membranes at different temperatures or as a calibration curve for finding the water potential of potato tissue using different sucrose concentrations (Figure 14.10).

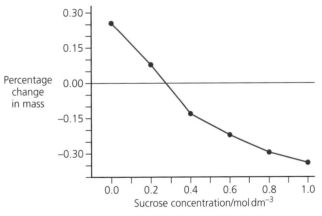

**Figure 14.10** Line graph showing the results of putting potato tissue in different sucrose solutions.

When you draw a line graph the independent variable is plotted on the $x$-axis, and the dependent variable is plotted on the $y$-axis. The axes should be labelled and you should include units. Someone should be able to look at the graph and know exactly what it shows without any further explanation. In the case of Figure 14.10, the $y$-axis has both positive and negative values because the potato tissue can gain and lose mass.

To choose a suitable scale, make sure all the points you need to plot will fit on the graph. Avoid a scale that involves parts of grid squares on the paper. Using whole squares is better. This makes it easier to plot points accurately. You should plot the individual points as clearly and accurately as possible. Use either a dot with a circle around it or a cross. You can draw either a smooth curve or straight lines joining the points. Your line should not extend beyond the first and last point you have plotted.

**TIPS**
- You should only draw a smooth curve if your data are sufficiently reliable for you to feel that you can confidently predict intermediate values. Otherwise, join the individual points with straight lines.
- Lines on line graphs are called curves even if they are straight lines.

Overall, the curve in Figure 14.10 is unusual in that it has an S-shaped or **sigmoidal** pattern. You should learn to recognise the other sorts of curves you might see in biology. Some curves are **linear**, others reach a **plateau** and some are **exponential**.

## Linear relationships

An example of a linear relationship is that between the rate of a reaction and substrate concentration when the enzyme is in excess. Provided the enzyme concentration remains in excess, the line will continue as a straight line. Graphs like this show a **linear relationship** which is described by the equation:

$$y = mx + c$$

where $m$ is the **gradient** and $c$ is the **intercept** on the y-axis. Figure 14.11 shows a line graph with a linear relationship like this, where the rate of reaction, $y$, at any point is equal to the enzyme concentration, $x$, multiplied by a constant, $m$, plus the value of the inttercept, $c$.

## Measuring rate of change

If the enzyme concentration becomes the limiting factor, the curve will reach a plateau (see Chapter 2). An example of this type of curve is shown in Figure 14.12.

Another line graph where the curve flattens off is shown in Figure 14.13. Here, note that the x-axis shows time rather than substrate concentration and the y-axis shows the concentration of the product rather than rate. At first, after an enzyme is added to some substrate, the reaction is fast because there is plenty of substrate to form enzyme–substrate complexes. Once some of the substrate has been used, the reaction begins to slow down.

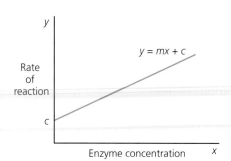

**Figure 14.11** Relationship between the rate of reaction and substrate concentration when the enzyme is in excess.

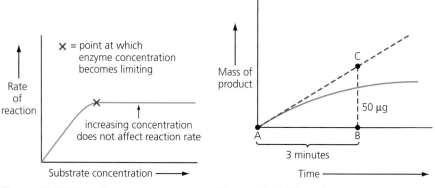

**Figure 14.12** When the enzyme concentration becomes the limiting factor, the curve flattens off.

**Figure 14.13** When the enzyme concentration becomes the limiting factor but with product concentration plotted against time.

The initial rate of the reaction, when it was going fastest, can be found by drawing a **tangent** at the start of the curve. A tangent is a straight line that matches the gradient on the curve. The rate can be found by measuring AB and BC and then calculating $\frac{BC}{AB}$. This gives the rate at which product is formed per minute. So if 50 µg of product were formed in 3 minutes, the rate would be $\frac{50}{3}$ = 16.7 µg/min.

### TIP

Draw a tangent by placing a ruler on a short part of the start of the curve. Start from the origin and match the gradient. Make sure the ruler is positioned carefully. A slight change can alter the position of the tangent a lot.

# Interpreting data from a sample of measured values

As you saw in Chapter 11, whenever a sample is taken from a larger group, the sample should be representative of the group as a whole. When this sort of data is collected, a large number of observations or measurements are made and so summarising the data is often helpful. Biological data from samples can be summarised in a variety of ways.

## Arithmetic means

One way of summarising sample data is by finding the **mean**. In Chapter 11, you saw that the mean is found by adding up all the measurements and then dividing this total by the number of measurements.

$$\text{Mean} = \frac{\text{sum of all measurements}}{\text{number of measurements}}$$

Table 14.5 shows the mass of a sample of 63 fish. The mean mass is the sum of all the masses divided by 63 which is $\frac{278.4}{63} = 4.42$ kg. Since the original masses are given with just one decimal place, the mean should also be given to one decimal place, in other words, 4.4 kg. The symbol for the mean is.

**Table 14.5** Mass in kilograms of a sample of 63 fish.

| 4.6 | 3.7 | 3.1 | 5 | 5.3 | 4.5 | 7.9 | 3.4 | 4.8 | 7.7 | 4.3 |
|-----|-----|-----|---|-----|-----|-----|-----|-----|-----|-----|
| 3.9 | 5.9 | 3 | 4.6 | 4 | 2.5 | 6.2 | 4.4 | 7.2 | 3.9 | 3.9 |
| 2.8 | 3.2 | 4.8 | 5.4 | 4.7 | 5.4 | 3 | 4 | 6.4 | 2.6 | 6.3 |
| 6.6 | 2.2 | 4.1 | 2.4 | 3.6 | 5.7 | 3.3 | 3.6 | 3 | 5.6 | |
| 4.2 | 3.2 | 2.1 | 6.3 | 3.3 | 3.8 | 5 | 5 | 3.5 | 3.3 | |
| 3.7 | 4.1 | 4.2 | 2.9 | 6.9 | 4.1 | 5.4 | 4.1 | 5.3 | 5.5 | |

## Standard deviation

The mean becomes more useful if a standard deviation is also given with it (see Chapter 11). Standard deviation is found from the formula

$$\text{SD} = \frac{\Sigma(x - \bar{x})^2}{(n - 1)}$$

where $\Sigma$ means 'the sum of', $\bar{x}$ is the difference between any measurement and the mean and $n$ is the number of measurements. To do the calculation, you need to find the difference between each measurement and the mean, square it and then add all the squared values together. Then divide this by one less than the number of measurements.

Standard deviation can be calculated manually, by using a scientific calculator or a by using a spreadsheet function. In this case, the standard deviation for the sample masses of fish shown in Table 14.5 is 1.36 kg. The mean mass can therefore be given as 4.4 ± 1.36 kg. This gives an indication of the spread of the data around the mean. Another set of fish masses with a smaller standard deviation would not show as much **dispersion** as this set of data.

## Range, median and mode

Other ways of summarising the data are the **range**, the **mode** and the **median**.

The range is the difference between the highest and lowest values. The highest value is 7.9 kg and the smallest 2.1 kg, so the range is 7.9 – 2.1 = 5.8 kg. It is another measure of dispersion.

The median is the middle value of the masses arranged in rank order. In this case, the median is 4.1 kg.

The mode is the mass that appears most often. In this sample, the mode is also 4.1 kg because it appears in the table four times more than any other value.

## Frequency tables and histograms

One way of summarising data is a frequency table. A frequency table is made by dividing the observations or measurements into a number of classes. **Classes** are either categories or ranges of measured values to which observations can be allocated. The **class frequency** or number of observations or measurements in each class can then be tallied up, as shown in Table 14.6.

**Table 14.6** Frequency table for the mass of a sample of fish.

| Mass/kg | Tally | Frequency |
|---|---|---|
| 2.0–2.9 | ///// // | 7 |
| 3.0–3.9 | ///// ///// ///// / | 16 |
| 4.0–4.9 | ///// ///// ///// //// | 19 |
| 5.0–5.9 | ///// ///// // | 12 |
| 6.0–6.9 | ///// / | 6 |
| 7.0–7.9 | /// | 3 |
| Total | 63 | 63 |

The data in Table 14.6 can be represented as a histogram. Figure 14.14 shows a histogram. The bars are plotted beside one another because the mass classes are continuous (they can take any value, within a range).

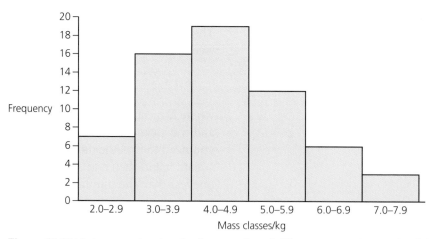

**Figure 14.14** Histogram showing the frequencies of different mass classes for fish.

A histogram is plotted when the frequency table shows measured data. Counted data in categories, such as the number fish of different species, would have to be plotted as a **bar chart**. In a bar chart, the bars are each separated by a space because the data is in discrete categories.

**TIP**
Make sure you know the difference between a histogram and a bar chart, and when it is appropriate to use each one!

# 15

# Developing practical skills

## Introduction

Alongside developing your knowledge and understanding of biology, you are expected to become familiar with a range of apparatus and competent in a number of practical techniques. You will gain experience of these apparatus and techniques during routine practical work. Your teacher will probably include a number of different practical investigations and activities in the course but there are some specified required practicals that you must complete. Written papers will assess your knowledge and understanding of the apparatus and techniques in some of these practicals. A list of the required practicals is given at the end of the chapter. But you can demonstrate your competence in these practical techniques during any of the practical work you do. Practical skills to be assessed in written papers are best covered by some of the questions in the other chapters.

## Investigating a scientific question

In all investigations you control something. This is the independent variable. In the oral rehydration solution investigation in Chapter 2, on enzymes, the independent variable is the concentration of amylase. As a result of these changes, you have to measure something else. This is the dependent variable. In this investigation, the dependent variable is the viscosity of the rice flour solution.

The first step in any investigation is to identify these two variables. The next is to plan how to change the independent variable.

### Changing the independent variable

An experimental result that occurs only once could be due to chance. It could also be because the scientist who carried out the investigation did it in a particular way. Experiments only provide **precise** data if they can be repeated by the same experimenter or by someone else to give very similar results. This means that you should give enough detail in describing your method for another scientist to follow the technique exactly as you intended, without the need for any help other than your set of instructions. In this case it is not enough, for example, to say, 'Put the mixture in a water bath'. Another scientist would want to know what temperature you used. 'Put the mixture in a water bath at a constant temperature of 35°C' is much better.

In most experiments, variables other than the one you are planning to change might also affect the results. We call these control variables. You should design your experiment so that these control variables are kept constant. There are several control variables in this experiment. The incubation time of the enzyme and rice flour solutions, temperature, pH and substrate concentration could all affect the rate of this reaction. If we keep these factors constant, then there is a reasonable chance that any difference in viscosity will be due to the concentration of amylase.

Once you have decided how you are going to change the independent variable and have identified the various control variables whose values you must keep constant, you need to ask another question: is there anything, other than the independent variable, in the way I have set up this experiment that could have produced these results?

If there is, you need a control. A control experiment is one that you set up to eliminate certain possibilities. In this investigation it is possible that the rice flour solution would have become runnier anyway if you just left it. We ought to have a control experiment where everything is the same as in the experimental container but without amylase or with denatured amylase. By setting up this control you eliminate the possibility that something other than amylase reduced the viscosity of the solution.

## Measuring the dependent variable

Once you have decided how you will change the independent variable, you should decide how to determine the effect of this on the dependent variable. You need quantitative data and this means taking measurements. These measurements must be made with the appropriate degree of **resolution**.

In the oral rehydration solution investigation, you could measure viscosity. For more resolution you could use a viscometer but it is not very likely that you would have one in your laboratory. There are other, simpler methods, however. You could pour the solution into a burette and see how long it took for the liquid to drain out through the tap, or you could take the time for a marble or other object to fall through a column of the solution.

You also need to bear in mind that if you carry out the experiment once only, you do not know if the results are precise. Carrying it out twice doesn't help you much either. You need to carry it out several times. Only in this way can you distinguish precise data from anomalies.

# Making quantitative measurements

A range of different quantitative measurements have to be made during practical work. Making and recording these measurements carefully is a key skill. The measurements you are likely to make will include time, volume, length, mass, temperature and pH. However, if you use instruments such as a colorimeter you will come across some less familiar measurements. A colorimeter measures the optical density of a solution. You may also come across less familiar units such as milliseconds (ms) or micrograms (µg).

Quantitative measurements
Measurements that involve numbers.

To enable some measurements to be made with more resolution, different apparatus needs to be used. A wall clock could be used to measure time in seconds but a stopwatch would be needed for measurements in milliseconds. Although you will be used to measuring length in millimetres you may not have measured length before in micrometres. You could measure the length of a leaf in millimetres using a **ruler**. But to measure the thickness of the leaf, you might use a piece of equipment called a micrometer and to measure cells or organelles seen with a microscope, a **graticule** would be used.

Sometimes a particular piece of equipment makes the measuring easier. A ruler would not be appropriate for measuring the length of a scallop shell because it is curved. **Callipers** would be more appropriate and would probably ensure a more **accurate** measurement to be made, even if the units being used were millimetres (Figure 15.1).

You may need more specialised instruments to make some measurements.

**TIP**

Do not use centimetres when measuring lengths. They are almost never a suitable unit to use in biology and can lead to decimal-point mistakes when converting to micrometres. Always use millimetres instead.

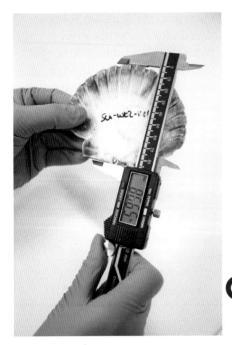

**Figure 15.1** Digital callipers being used to measure shells.

The pH of a solution is measured using a pH meter (Figure 15.2). A **potometer** measures the volume of water taken up by a leafy plant stem. A **respirometer** measures the volume of gases involved in gas exchange. A **colorimeter** measures how much light can pass through a coloured solution so one can be used to compare how much of a purple pigment there is in solutions containing different samples of beetroot tissue. Figure 15.3 shows a colorimeter and a potometer.

You may use either of the two pieces of equipment in Figure 15.3. If you do, you will need to know about the problems you may encounter when using them and how to avoid or overcome them.

Colorimeters often need to be zeroed, rather like a balance, before making measurements. Zeroing a balance is done with

**Figure 15.2** A pH meter being used to measure pH of a solution.

**Figure 15.3** (a) A colorimeter measures optical density and (b) a potometer measures the volume of water taken up by a leafy plant shoot.

nothing on the balance, but a colorimeter is zeroed using a clear solution, often water, called a **blank**. Samples are placed into a colorimeter in rectangular plastic tubes called cuvettes. Just as it is important not to spill material or solutions onto a balance pan to avoid errors, it is important to avoid spilling drops of liquid on the outside of the cuvette. This is because drops of liquid can cause errors by scattering the light. Leaks at the joints of a potometer or water on the leaf surfaces can cause errors in the volume measurements. Knowing and avoiding likely causes of error with each piece of equipment helps to ensure accurate measurements.

> **TIP**
> Put the blank back into the colorimeter every so often between measurements and check that it still reads zero. If not, re-zero it. This ensures that all of your measurements remain as accurate as possible.

Most colorimeters have a method of placing coloured filters in the path of the light. This is because light of a particular wavelength is more suitable for measuring the optical density of solutions of particular colours. For example, green light is most suitable for measuring the optical density of purple solutions from beetroot tissue. In general, the filter that gives the highest optical density reading for a solution of a certain colour is the best one to use.

Potometers work by measuring how far an air bubble travels along a scale in a certain time. It is important to avoid other air bubbles elsewhere in the apparatus, so they are often assembled underwater in a large sink or bowl. Once they are working without leaking, it is best not to disturb them. Potometers often have a three-way tap and a water reservoir to enable the air bubble to be returned to the start of the tube by letting water back in without having to remove and replace the stem. This enables repeat readings to be made easily.

> **TIP**
> Keep the leaves of your leafy stem out of the water while assembling the potometer in the sink. Any water on the leaf surfaces will interfere with the transpiration rate and cause an error in measuring the volume of water taken up by the stem and it is very difficult to dry them without damaging them.`

## Using glassware

Measuring volumes accurately using glassware such as measuring cylinders is a basic practical skill. A common requirement in many practical investigations is a series of concentrations of a given

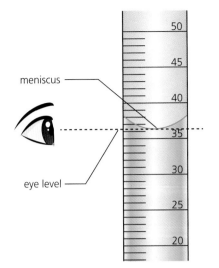

meniscus

eye level

**Figure 15.4** How to read a volume correctly from a meniscus.

solution, such as using a series of substrate concentrations to see how this affects the activity of an enzyme. Selecting the appropriate size of measuring cylinder depending on the volume to be measured is crucial. The smaller the graduations on the measuring cylinder, the better the resolution of the measurement can be. But obviously the total volume has to be able to fit into the measuring cylinder selected.

Reading the position of the **meniscus** is also vital for accurate volume measurement. The correct way to do this is to line up the bottom of the meniscus with the graduation mark, at eye level, as shown in Figure 15.4.

Ensuring that the full volume is drained out of the measuring cylinder is also important, as is the need to avoid splashes and spillages, which lead to error.

# Using an optical microscope

The optical microscope is a vital tool for biologists because of the small size of many biological structures. Tissues, cells and organelles are all too small to be seen in any detail by eye so their images must be magnified. Optical microscopes work by directing light through a thin layer of biological material supported on a glass slide. This light is then focused through several lenses so an image can be seen through an eyepiece. You can switch from low to high power by rotating a different objective lens into position.

Optical microscopes all have the same basic design. Figure 15.5 (overleaf) shows a typical optical microscope.

Any material you are going to look at using an optical microscope has to be either transparent or really thin to allow light to pass through it. Some material might already be thin enough, such as the epidermis peeled from a leaf to look at stomata or some lamellae removed from a piece of fish gill. Other material such as lung tissue or leaf tissue may need to be sliced very thinly using a sharp blade. Root tip tissue can be softened and squashed under the cover slip so it is spread out into a thin layer. On the other hand, if you want to look at live water fleas to measure their heart rate they obviously cannot be squashed into a thin layer. Fortunately, they are quite transparent animals so this allows them to be observed alive using an optical microscope.

**TIP**
Even if you are going to look at your material under high power, always start by focusing on low power to avoid damage to high-power lenses. It is easier to see what you are looking at using low power and when you change to the high-power objective it should be roughly focused already.

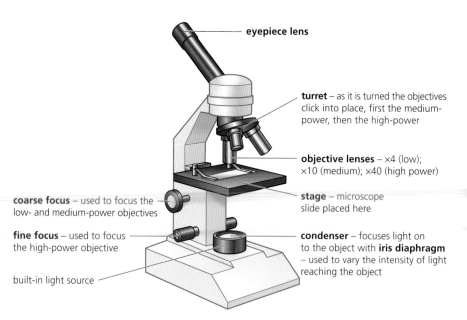

eyepiece lens

**turret** – as it is turned the objectives click into place, first the medium-power, then the high-power

**objective lenses** – ×4 (low); ×10 (medium); ×40 (high power)

**coarse focus** – used to focus the low- and medium-power objectives

**stage** – microscope slide placed here

**fine focus** – used to focus the high-power objective

**condenser** – focuses light on to the object with **iris diaphragm** – used to vary the intensity of light reaching the object

built-in light source

**Figure 15.5** A typical optical microscope showing the main components.

Thin layers of material will quickly dry out and shrivel up in the heat of the microscope lamp. This is why, in the case of leaf or gill tissue, a drop or two of water is usually added beneath the cover slip to prevent dehydration damaging the cells. Animal cells such as blood cells need a drop or two of solution of sugar or salt adding instead. Most animal cells will also be damaged by osmosis bursting them if just water is added to the slide.

Some animal and plants cells such as red blood cells or parts of them, such as the chloroplasts in leaf cells, are coloured and easy to see. If not, you may need to add a **stain** so that the structures you wish to see are visible. This is how chromosomes can be seen in transparent root cells undergoing mitosis. A stain such as acetic-orcein is added which stains DNA. Any chromosomes that have shortened and thickened sufficiently are coloured by the stain and enable any cells in different stages of mitosis to be identified.

**Isotonic solution** A solution with the same water potential as the cell cytoplasm.

## TIPS

● If you are cutting material with a sharp blade to make a thin section, do not just push the blade downwards because it often squashes the material before it cuts it, damaging the tissues. Also make sure that you cut away from you and onto a suitable hard surface. Move the blade back and forth gently as if you were sawing through a piece of wood.
● If you add too much stain to material on a slide, you will not be able to see anything except the stain. Always add a small volume. More can always be drawn under the cover slip by adding a drop on one side and touching a piece of tissue at the other. The stain is pulled under the cover slip.

If you need to make measurements of cells seen using an optical microscope, then a graticule is required. This provides a scale like a ruler in the field of view. It has no fixed units, so it has to be **calibrated** for the objective lens being used. Calibration is done using a stage micrometer, which is a second scale in micrometres engraved on a microscope slide. Using the two together, as in Figure 15.6, it is possible to see how many micrometres each graticule 'unit' is worth. If you continue to use the same objective lens, the graticule can now be used like a ruler.

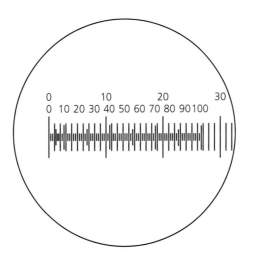

**Figure 15.6** Graticule and stage micrometer being used together to calibrate the graticule.

# Making drawings

Making drawings is an important way of recording the results of practical work in biology because often the outcome of an investigation is descriptive rather than quantitative. A record of what you have done is more easily made as a drawing. The purpose of biological drawing is to make a clear scientific record of what you have observed rather than an artistic interpretation of the material. This means line drawings in pencil with no shading and no use of colour.

The important things are shape, proportion and scale. The structures drawn should be the same shape as those you observe, the parts of the drawing should be in proportion to one another and the drawing should be large enough to show the details clearly. If you are drawing individual cells, just draw two or three cells as examples, perhaps to show the range of shapes or sizes. If you are drawing tissues such as those in a leaf section or in the wall of a blood vessel, individual cells should not be shown. Instead, a **tissue map** should just outline the areas of different tissue, as shown in Figure 15.7 (overleaf). If you are drawing an organ such as a heart, there is no need to attempt to show three dimensions by shading. A **line diagram** is all that is required, as shown in Figure 15.8 (overleaf).

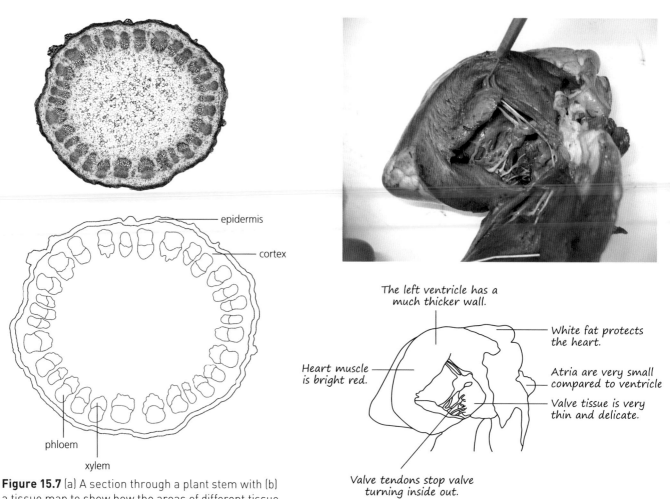

**Figure 15.7** (a) A section through a plant stem with (b) a tissue map to show how the areas of different tissue should be drawn.

**Figure 15.8** (a) A sheep heart together with (b) an annotated drawing.

Annotating drawings makes them much more useful. **Annotations** are labels with more information than just the name of the structure. Annotations are where you can record the colours, textures and other properties of the things you have observed. You might include the name of the tissue or structure, but equally, annotations can just be descriptive or they might include measurements. Examples of annotation are shown in Figure 15.8. For clarity you should try to avoid label lines crossing each other or too much of the drawing. Label lines should be straight, drawn in pencil, using a ruler.

**TIP**

All drawings should have a title and some indication of scale or magnification. If drawing from a microscope, the drawing should show the magnification used to observe the material.

# Identifying biological molecules

Being able to identify which biological molecules are present in material from organisms is a useful part of investigating how they work. There are some simple chemical tests that can identify some of the biological molecules that are present in materials or solutions (see Table 15.1 and Figure 15.9).

**Table 15.1** Four simple chemical tests for biological molecules and their positive results.

| Test for | Reagent | Initial colour | Positive result |
|----------|---------|----------------|-----------------|
| Protein | Biuret reagent | Pale blue | Violet solution |
| Reducing sugar | Benedict's reagent | Blue | Red precipitate |
| Starch | Iodine solution | Orange | Dark-blue solution |
| Lipid | Ethanol | Colourless | Cloudy suspension |

**Figure 15.9** Positive results for the four chemical tests.

(a) Positive result for a lipid test.

(b) Benedict's reagent (left) and positive result for a reducing sugar test (right).

(c) Positive result for a starch test (left) and positive result for a protein test (right).

**TIP**
You need to learn both the colour of the test **reagent** and the colour of a positive result.

Biuret reagent and iodine solution are simply added to the sample. A few drops are all that is needed. The colour change showing a positive result becomes apparent immediately if either protein or starch is present (Figure 15.9c). Benedict's reagent must be heated for a few minutes with the sample to show a positive result. A few drops heated to 80°C for several minutes will give a positive result if reducing sugar such as glucose or maltose is present (Figure 15.9b). The test for lipid is known as the emulsion test. A volume of ethanol equal to the volume of the sample is added and the mixture shaken well. The mixture is then poured carefully into water. If lipid is present, a cloudy white suspension is formed (Figure 15.9a).

**TIP**
The positive result for the reducing sugar test can be confusing. Small concentrations of reducing sugar result in small quantities of red precipitate, which can look orange or even green in the blue solution. All of these are positive results, but they indicate different concentrations of reducing sugar.

Benedict's reagent can also be used to test indirectly for non-reducing sugars. If the sample gives a negative result when tested but a non-reducing sugar such as sucrose might be in the sample, then add a few drops of acid, warm and repeat the test. A positive result the second time around indicates that non-reducing sugar was present.

## Separating biological molecules

Chromatography is a way of separating unknown mixtures of biological molecules and is carried out on paper or on thin layers of solid media such as alumina on glass or plastic plates. The mixture of compounds, perhaps a mixture of unknown sugars or amino acids, is very carefully pipetted onto an **origin** line one small drop at a time (Figure 15.10). In between each drop, the spot is dried. A small but concentrated spot is slowly built up on the origin line with repeated addition of drops and drying between each one. Some 20 or 30 small spots can be applied to the one spot.

**Figure 15.10** Spotting solutions onto the origin line of a thin-layer chromatography plate.

A series of spots of **standards** are added along the origin line. These are solutions of just one sugar, such as glucose or fructose, or amino acids such as glycine or alanine. The standards and the mixture spots are labelled or numbered.

When the mixture and the standard have been spotted onto the origin line, the plate or paper is held vertically in a small volume of chromatography solvent. The solvent slowly moves up the paper or the solid medium on the plate. This may take several hours. As the solvent moves, the biological compounds are carried up with it. But they do not all move at the same rate. Those that move faster travel further up the plate or paper. So a mixture spreads out into a series of individual spots. The standards can then be compared in terms of their positions with the spots in the unknown mixture, enabling the components of the mixture to be identified.

**TIP**
Label your spots with pencil rather than pen and beneath the origin line. Pencil will not run when the plate or the paper is added to the solvent and the labels will be out of the way below the line.

**TIP**
Check that your solvent is shallow enough to remain below the height of the origin line when you stand your plate or paper in it. Otherwise your spots will simply be washed off into the solvent.

# Using living organisms

If you use live animals such as water fleas (*Daphnia*) to observe heart rate or insects such as locusts to look at spiracles, you should take great care to avoid them being harmed. They should only be observed for a short period of time and returned to a larger container as soon as possible. Water fleas should be placed on **cavity slides** so sufficient water can be provided on the slide to protect them from the heat of the microscope lamp and dehydration.

If you carry out investigations on other members of your class, for example on breathing rate or pulse rate before and after some exercise, you should be very careful to avoid injury. The exercise should be planned carefully and done in a suitable location to prevent accidents.

**TIP**
Turn the microscope lamp off whenever possible to avoid heating your water fleas too much.

## Growing microbes

Growing bacteria successfully requires **sterile conditions** so that the culture is not contaminated by other microorganisms that you are not investigating, especially potential human pathogens. Appropriate procedures to ensure aseptic technique should be followed.

**Aseptic techniques** are ways of working that minimise the possibility of contamination or escape of your bacterial culture. Before starting any work with bacteria, the work surface should be sterilised using a suitable disinfectant. Any equipment such as wire loops, pipettes or spreaders that will come into contact with the bacteria should be sterilised before use.

It is a sensible precaution to close any nearby windows or doors to prevent draughts that might carry bacteria towards or away from your working area. You should have everything close to hand before starting any transfers of bacteria. Once you begin, you should work quickly but carefully. Bottles or Petri dishes containing bacterial cultures should only be opened for the shortest time possible to reduce the chance of any bacteria getting in or out.

Whenever you open a culture bottle or test tube to remove bacteria, you should immediately move the neck of the container into and out of the Bunsen burner flame while holding the container almost horizontally. This is known as **flaming**. It sterilises the neck of the container and the air above the culture. Before opening the bottle, you should also flame the wire loop you intend using by holding it in the flame until it becomes red hot.

Whenever you open a Petri dish you should hold the lid at an angle above it while you make any transfers as quickly as possible. The lid should then be replaced immediately. Any used pipettes or spreaders should be placed into a suitable disinfectant.

Testing for antibiotic resistance usually involves growing bacteria on nutrient agar (Figure 15.11) and then adding antibiotic on prepared discs of filter paper to the surface of the agar. Where bacteria have been killed by the antibiotic, clear zones are formed around the disc.

Bacteria for plating out are usually cultured in a nutrient broth first. This is a liquid growth medium rather than a solid one. Once the culture has grown,

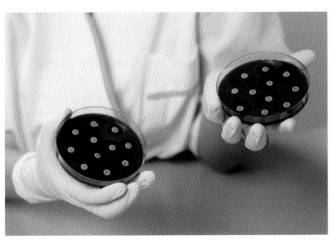

**Figure 15.11** Bacteria growing on nutrient agar in Petri dishes, being tested for antibiotic resistance.

it is plated out by spreading some over the surface of an agar gel. The effectiveness of antibiotics on the bacteria is much easier to see on an agar plate rather than in broth.

## Dissecting animal or plant organs

Observing biological material frequently requires that it be cut open or sectioned so the internal structures are visible. To see the internal features of a heart such as the relative wall thicknesses and the valves requires the ventricles to be cut open. To see the distribution of xylem and phloem in a plant stem, thin cross-sections have to be cut. Sections of gill may need to be removed from a fish such as a herring and the very ends of garlic roots removed for making root-tip squashes to see mitosis.

These activities require the use of sharp instruments such as scissors and scalpels. Handling these confidently but safely is an important practical skill in biology, as is being careful enough to cut and slice the material without damaging it so much that observations are impossible.

Cutting should always be done away from you and your fingers, and on a suitable surface to avoid damage to the work bench. Sharp instruments are better than slightly blunt ones because they require less force to cut and there is therefore less risk of them slipping suddenly.

# Using sampling techniques

Investigating variation in populations such as dog whelk shell length or plant leaf area usually requires a sample of the population to be measured. It is important that the sample is random so as to avoid any bias introduced by deliberate selection of individuals because they are easier to spot or easier to collect or for any other reason.

Collecting animals such as banded snails from a hedge or dog whelks on a rocky shore requires that individuals simply be picked up. But to ensure a random sample, a quadrat should be placed at locations determined by randomly generated co-ordinates and a certain number, say five, collected from each of a number of randomly positioned quadrats. Co-ordinates can be generated by rolling dice, by pulling numbers from a pocket, using random number tables or a random number generator on a computer.

Collecting leaves for measuring variation requires a different approach. Leaf area is partly dependent on age of the leaf, so a valid sample should attempt to include only leaves of the same age. Counting back from the shoot tip to, say, the third or fifth leaf back on the stem each time ensures that the leaves are all of comparable age. A number of plants can have their leaves labelled around the leaf stem with a sticky label and a number. Again, rolling dice will generate random numbers which can be used to select a sample of leaves for measurement without any bias.

Selecting random areas of a leaf surface under the microscope for finding the density of stomata is a form of sampling too. A slide with an engraved grid (Figure 15.12) can be used and stomata counted within squares determined once again by random numbers or co-ordinates. The area of the grid squares or the area of the field of view can be found by making measurement using a graticule.

**Figure 15.12** Using grid squares as random samples to find the density of stomata.

## Using digital technology

Measuring human pulse rate can be done with a pulse sensor linked to a datalogger that will record the data for a specified time period and enable graphs to be produced very easily. When counting your own pulse rate, or that of a partner, it is easy to miscount. A pulse sensor avoids this problem.

Variation data entered into a Microsoft Excel spreadsheet allows means and standard deviations for each set of data to be calculated and compared very quickly and conveniently. The data such as dog whelk shell length is added to a column and then AVERAGE and STDEV functions can be used to find the mean and the standard deviation for data in a range of cells.

Animations that model the effects of natural selection such as predation are very helpful for illustrating how allele frequencies will change over generations if mutation gives rise to an advantageous allele. The advantage of an animation is that it can model the events over many generations in a short space of time.

## Required practical activities

1 Investigation into the effect of a named variable on the rate of an enzyme controlled reaction (see Chapter 2)

2 Preparation of stained squashes of cells from plant root tips; set-up and use of an optical microscope to identify the stages of mitosis in these squashes and calculation of a mitotic index (see Chapter 5)

3 Production of a dilution series of a solute to produce a calibration curve with which to identify the water potential of plant tissue (see Chapter 3)

4 Investigation into the effect of a named variable on the permeability of cell-surface membranes (see Chapter 3)

5 Dissection of animal or plant respiratory system or mass transport system or of organ within such a system (see Chapter 9)

6 Use of aseptic techniques to investigate the effect of antimicrobial substances on microbial growth (see Chapter 11)

**TIP**
Processing and presenting data are important skills. Make sure you know about these and other skills developed in the required practicals.

265

# 16 Exam preparation

## Overview

Your exam preparation begins from the moment you enter your first A-level biology lesson. The most successful candidates drive their own learning and do not rely entirely on their teacher or the work they do in lessons. The exams include an element of recall, but simply knowing the facts will not get you a good grade. Most of the marks require a demonstration of understanding and the ability to apply your knowledge in a new context. It is important that you are thoroughly familiar with the topics covered, but also that you have a good biological general knowledge. If something does not make sense, or some information seems to be missing, you should investigate on your own account using the internet and any available books. Scientific magazines and TV programmes should also be used wherever possible to broaden your knowledge base. This will bring you into contact with information, examples and new contexts, which will reinforce your knowledge and understanding of the work on the A-level specification. The BBC News website also has interesting articles in the science section.

A thorough familiarity with the AQA specification is vital. This will tell you precisely what you need to know (and therefore what you do not need to know) and gives details of the assessment process. You will need to be familiar with the format of the exam papers and practical assessment.

If you constantly 'read around' the subject as you go through the course, you will need to spend less time on revision, as many of the basic facts will be hardwired into your brain. Nevertheless, thorough and effective revision will be necessary. If you have a thorough strategy or plan for your revision, you can improve your grades.

Once you get into the exam, the mark you achieve will not depend only on how much you know and understand about the subject. Exam technique plays a big part in the final outcome, as it is essential that you answer precisely what the questions ask and explain yourself clearly. You cannot get marks for material you do not know or understand, but it is essential that you gain every possible mark from the content that you do know.

This chapter outlines some links and tips which will guide you in the right direction. However, different people do learn in different ways and you will need to find out the working practices and revision techniques that work best for you.

# The AS examination

The structure of the AS examination is shown in Figure 16.1.

| Paper 1 | Paper 2 |
| --- | --- |
| Written exam 1 h 30 min | Written exam 1 h 30 min |
| Any AS content, including relevant practical skills | Any AS content, including relevant practical skills |
| 75 marks | 75 marks |
| Worth 50% of AS | Worth 50% of AS |
| Short-answer questions (total 65 marks) | Short-answer questions (total 65 marks) |
| Questions on a short comprehension passage (10 marks) | Structured continuous prose question (10 marks) |

(Paper 1) + (Paper 2)

**Figure 16.1** The structure of the AS examination.

The exam will assess different skills. There are three different assessment objectives, as follows.

A01 Demonstrate knowledge and understanding of scientific ideas, processes, techniques and procedures.

A02 Apply knowledge and understanding of scientific ideas, processes, techniques and procedures:
- in a theoretical context
- in a practical context
- when handling qualitative data
- when handling quantitative data.

A03 Analyse, interpret and evaluate scientific information, ideas and evidence, including in relation to issues, to:
- make judgements and reach conclusions
- develop and refine practical design and procedures.

These skills will be assessed in the two AS papers as shown in Table 16.1.

**Table 16.1** Skills to be assessed in the two AS papers.

| Assessment objective | Component weightings (approximate %) | | Overall weighting (approximate %) |
| --- | --- | --- | --- |
| | Paper 1 | Paper 2 | |
| A01 | 47–51 | 33–37 | 35–40 |
| A02 | 35–39 | 41–45 | 40–45 |
| A03 | 13–17 | 21–25 | 20–25 |
| Overall weighting of components (approximate %) | 50 | 50 | 100 |

A01 questions are the ones that test whether you can understand and recall what you have learned. Notice that this kind of question accounts for less than half the marks overall, so just learning your notes is not enough to get you a reasonable grade in the exam. A02 questions check that you have an in-depth understanding of what you have learned, by presenting you with an example or context that you haven't seen before, and asking you to apply your understanding in a new situation. This means you really need to understand thoroughly what you have learned. These questions account for almost half the marks. Finally AO3 questions ask you to evaluate and analyse scientific information or data or relate to practical skills. These are worth up to a quarter of all marks.

In summary, if you hope to achieve a high grade you really need to understand everything inside out, and be good at thinking scientifically.

Table 16.2 presents a list of key words that are used in exam questions, with explanations of what examiners mean when they use each term.

**Table 16.2** Key words used in exams.

| Describe | This simply means 'tell us about…'. Or 'give an account of'. This is a straightforward test of what you have learned and remember. Alternatively, you might be asked to summarise the trends shown in a graph or table of data that you are presented with. |
|---|---|
| Explain | This means 'give one or more reasons for'. The best way to make sure that you explain, rather than describe, is to use the word 'because' in your answer. For example, 'This happens because…'. |
| Suggest | This word is a clue that you are being given an unfamiliar example or context, but you are being asked to apply your AS level understanding in a new or unfamiliar situation. The answer to this question is unlikely to be in your notes, but you should have enough understanding of biology to work out the answer |
| Evaluate | This means you need to present arguments for and against a point of view or conclusion. A good way to answer questions like this is to think of evidence to support the statement or idea, and then think of evidence why the statement or idea might not be supported. Alternatively, it may be asking you to make a judgement about something, such as an experimental protocol. |
| Give | Produce an answer from recall or from given information. Don't explain. |
| Sketch | This means 'draw approximately'. For example, if you are asked to sketch a line on a graph to show what would happen in different circumstances from the line already on the graph, the examiner is only looking to see that you have the right trend and shape on the line you draw. The examiner does not expect mathematical accuracy. |

# Different types of question

Your exam will consist of several different kinds of question. You should look at past papers and specimen papers so that you recognise the different kinds of question, and understand the best way to answer them.

## Structured questions

There are two kinds of structured question. Some require short answers and others require longer answers.

Short structured questions typically score one or two marks. Sometimes only a single word is needed, or a simple calculation.

Longer structured questions typically score three or four marks, although they could be up to six marks. Sometimes you are asked to use information from a table, graph or diagram in your answer, which you must do. Sometimes you are given two command words, such as 'Describe and explain the mechanism that causes forced expiration'. To answer the question well, make sure that you describe and explain. The mark scheme for this question has four marking points:

1 contraction of internal intercostal muscles

2 relaxation of diaphragm muscles/of external intercostal muscles

3 causes decrease in volume of chest/thoracic cavity

4 air pushed down pressure gradient.

Notice how the first two marking points describe the mechanism that causes forced expiration. The second two marking points explain how these actions bring about forced expiration.

Sometimes structured questions involve reading some information at the start of the question. Don't be tempted to skip straight to the questions without reading the information you are given very carefully. Examiners do not give you information to read unless it is necessary to answer the question.

## Comprehension questions

In these questions, you will be asked to read a short passage. You should read the passage carefully, then read through the questions that follow. Often, the questions that follow will tell you which lines in the passage the questions are referring to. After you have read all the questions, read the passage again and then answer the questions, one by one. You may be asked to use your own knowledge as well as information from the passage to answer the questions. These questions are testing whether you can relate the biology you have learned in your first year's study to a new situation.

## Extended prose answers

These questions typically score five marks or so. These require you to answer clearly and in a well-organised fashion. For example, you may be asked 'describe the differences between active and passive immunity (5 marks)'. The mark scheme may have more than five marking points on it, so you can get full marks by mentioning just five of them. However, be careful that in this example you actually point out the differences. If you simply describe active immunity, and then passive immunity, you are not pointing out the differences. The examiner will not do the work for you. Here is the mark scheme for this question:

1 active involves memory cells, passive does not

2 active involves production of antibody by plasma cells/memory cells

3 passive involves antibody introduced into body from outside/named source

4 active is long term, because an antibody produced in response to antigen

5 passive is short term, because an antibody (given) is broken down

6 active (can) take time to develop/work, passive is fast acting.

Take a look at AQA's website for more sample questions and mark schemes. This will give you a good idea how marks are allocated to sample questions.

Some extended prose questions simply ask you to describe a process, for example: 'When a vaccine is given to a person, it leads to the production of antibodies against a disease-causing organism. Describe how. (5 marks)'. Here, you need to give a clear account of the process, but make sure the points are expressed clearly and in the correct sequence. Good use of scientific terminology is also important.

In this example, the mark scheme has seven marking points and you only need five of them:

1 vaccine contains antigen from pathogen

2 macrophage presents antigen on its surface

3 T cell with complementary receptor protein binds to antigen

4 T cell stimulates B cell

5 (with) complementary receptor on its surface

6 B cell secretes large amounts of antibody

7 B cell divides to form a clone all secreting/producing the same antibody.

## Synoptic questions

In answering questions you will be expected to use knowledge and understanding from all the aspects of biology you have studied this year. For example, you may be expected to refer to the primary, secondary or tertiary structure of proteins when answering questions about enzymes and explaining their specificity. Similarly, you may be expected to discuss the specificity of proteins when answering a question about a test that uses monoclonal antibodies.

# Answering data questions

## Describing a graph

- This means give the trend shown by the graph: don't describe every single point.
- A-level examiners will never give you a simple straight line, so saying 'when x goes up, so does y' is very unlikely to be the answer required.
- There is usually a change in the gradient of the graph, so make sure you notice this. For example, the answer might be, 'As x increases from … to …, y increases slowly, but after x reaches the value of …, y increases much more steeply'.
- Examiners may give you a graph showing a correlation, but this may not be exact. You may be asked to suggest a reason why there is a trend but the correlation is not exact.
- Remember that a correlation between two different factors does not mean that one causes another.

As an example, describe the graphs shown in Figure 16.2.

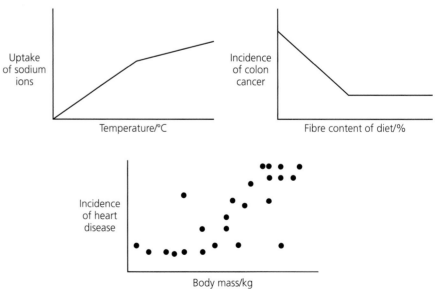

**Figure 16.2**

For example, a description of the first graph would be something like 'As the temperature increases from … to …°C, the rate of sodium ion uptake increases. As temperature increases above …°C, the rate of sodium ion uptake continues to increase, but more slowly'.

If you are then asked to **explain** the data on a graph, you need to give a reason for what you see. Your answer would start 'This happens because…'.

## Evaluating experimental designs

- Having a large number of samples or people in an investigation makes the results more **reliable** because any outliers or unusual cases have less of an effect on the overall results.
- We often present results as **percentages** so that we can **compare** the results from two different groups more easily, especially if there are different numbers in the groups.
- A good investigation changes only one variable. For example, if you want to find the effect of temperature on the rate of an enzyme-controlled reaction, what variable do you change and what variables do you keep the same?
- In human studies, we usually can't control all the variables. Therefore, the results might not be completely reliable. This would be the case in a study to find out whether smoking cigarettes increases the chances of developing heart disease.
- It is easier to control all the variables if you do an investigation on animals, but remember that studies on animals might not be applicable to humans because the animals' physiology may be different from ours.
- If you are testing a drug, you usually give each person a dose calculated in milligrams per kilogram of body mass, rather than just giving everybody the same dose of the drug. Why?
- Good investigations have an experimental group and a control group. For example, if you are finding the effect of a probiotic yogurt drink on the incidence of colon cancer, you would have your experimental group taking a probiotic yogurt drink every day. What would your control group be doing? And why?
- Remember that the control group is treated in **exactly the same way** as the experimental group, except for the one factor that you are changing.

Sometimes results are given like those in Table 16.3, with standard deviations.

**Table 16.3** An investigation to compare loss in body mass over 4 weeks on a low-glycaemic-index (GI) diet and a normal reduced-energy diet.

| Group | Mean loss in body mass/kg (± standard deviation) |
|---|---|
| Normal reduced-energy diet | 1.7 ± 0.1 |
| Low GI diet | 1.9 ± 0.3 |

- How would you set up an investigation like this?
- Why did the investigator find a mean loss in body mass for each group?
- Do these data show that the low-GI diet leads to greater weight loss? Explain your answer.
- What does standard deviation show?

## Calculations

Make sure you show your working. At AS level, a calculation is often worth two marks, but one mark can be awarded if your method was correct but the answer was wrong.

# Exam technique

- Read the question carefully and ensure your answer is accurately targeted to what it asks. If you look at mark schemes you will see that the marking points are very specific and correct but irrelevant points will not be credited.
- Your exams are worth 75 marks and the paper takes 90 minutes. Calculate in advance of the exam how much time you will need to read the questions and then check your answers at the end. Divide the remaining time between the marks available to allocate the time you want to spend on each mark during the exam. Use the number of marks for a question to ensure you have written enough, but not too much. A three-mark question needs three valid points to be made, for example.
- Question writers do not include information that has no use. If you get to the end of the question and there is some information you have not used, alarm bells should ring in your head. If you get stuck on a question, look carefully through the question for 'clues' to the answer.
- Communication must always be clear and detailed. Use appropriate scientific terminology. Avoid vague and general comments which do not include factual information. If you find yourself writing something that you haven't learned during the first year of study of your biology course, then think carefully about whether this is the answer the examiners are looking for.
- Never leave a question blank, especially if it is worth three marks or more. If you find it difficult, leave it and do an easier question, but come back to it at the end of the exam and have a go.
- Don't waste time writing out the question again. If the question is 'Name the monomer from which proteins are made' you just need to write 'amino acids'. Don't waste time with 'The monomers from which proteins are made are amino acids'.
- Make sure you read the whole question before starting to answer. There may be useful information in it, or key words to help you answer the question precisely.
- If you are asked for two things, such as two reasons for something, don't give three. The examiners will only mark the first two answers that you give. If there are more than two answers, and one or more of them is wrong, the examiners will use the wrong answers to cancel out a correct mark earlier in the list.
- Be clear what you are referring to in your answer. Avoid starting the answer with 'It' or 'This'. Make it clear what you are referring to; for example, 'The enzyme…'.
- Write clearly and legibly using a **black pen**. This is because your exam paper will be scanned so it can be marked on a computer. Don't write your answer in the margin or in a space at the bottom of a page. If you need more space, ask for an additional sheet of paper and write the question number clearly before the answer.

# Communication skills

Improve these answers, all of which are factually correct but lack clarity or detail.

1 The folded membrane of a mitochondrion is an adaptation for more efficient respiration.

2 The thin wall of the alveolus allows for easier exchange of gases.

3 The reason for the higher blood pressure in the left ventricle is because it needs to pump blood all around the body.

4 The Benedict's test is the test for sugars. The test solution has Benedict's solution added and is then heated. If sugar is present, there is a colour change to brick red.

## Revision

To revise effectively, it is useful to have a basic understanding of how memory works. When you receive information, either for the first time or when you are revising something that you have completely forgotten, it enters your working (or short-term) memory. From the point of view of passing exams, working memory is not very useful. It fades very quickly, and if you are going to have any hope of remembering what you've learned, you have got to get it into your long-term memory. Fortunately, this is not difficult. All you really have to do is pay close attention to the information as it comes in, and try to make some sense of it. Information that is meaningless to you will not make it into your long-term memory. It is therefore important that you ask questions of your teacher (or a computer search engine) if you do not understand the information you are given.

However, getting the information into your long-term memory is only half the battle. It is no good having it stored somewhere in your brain if you cannot retrieve it again.

Evidence suggests that the information will stay in your brain for quite a long time, possibly forever, but your ability to access it may fade. Effective revision methods will optimise the retrieval of information, but you will always forget a good proportion of what you have heard or read (see Figure 16.3).

**Figure 16.3** Percentage of information retained after different time intervals.

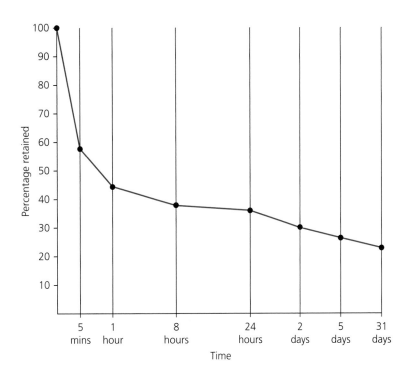

This is not quite as bad as it seems: a lot of what you hear or read is not important. For example, you need to retain the basic ideas, but not the actual sentences used word for word. There are a number of established learning techniques that will greatly increase how much you remember.

- **Repetition**: review new or revised material a week or so after encountering it. This need only be a brief refresh of the memory: you do not have to learn the material all over again. Try to do regular reviews as you go through the course, as this will mean that you already remember much of the material when you start intensive revision.
- **Forming memory cues**: link the information with something you already know. This may be other parts of the course, or personal experiences. The human aspects of biology are particularly suited to this: you have experienced things like breathing, sweating and reflexes, you may know someone with diabetes, etc.
- **Mnemonics**: a mnemonic is a formula or rhyme which assists the memory. A common form is to use the initial letters of a list of words to form a memorable sentence. For example, when memorising the stages of mitosis (prophase, metaphase, anaphase, telophase) you could use the phrase 'penguins march around trees'. When doing this, it is important that the sentence is very memorable: it is better to choose things that are visual and a bit ridiculous! The mnemonic forms a memory cue, but this type of cue is mainly restricted to lists.
- **Translate the information into a different form**: use the information to make a mind map, a picture or a presentation, or explain what you have learned to someone who knows nothing about A-level biology. If you cannot make someone else understand the information, you may not fully understand it yourself.
- **Do not revise for long without a break**: most peoples' brains cannot take in information continuously for longer than about half an hour without a break. A short break of a few minutes will refresh your brain and allow you to revise more efficiently.
- **Structure your revision**: it has been established that people learn best at the beginning and the end of a session. Start your revision with material that is quite difficult to understand or remember, finish it with similar material (or a review of what you learned at the start) and do the easier content in the middle. This is also a reason for breaks in your session: with breaks, you get more 'beginnings' and 'endings'.

# Index

# Free online resources

Answers for the following features found in this book are available online:

- Prior knowledge questions
- Test yourself questions
- Activities

You'll also find an Extended glossary to help you learn the key terms and formulae you'll need in your exam.

Scan the QR codes below for each chapter.

Alternatively, you can browse through all chapters at:
www.hoddereducation.co.uk/AQAABiology1

## How to use the QR codes

To use the QR codes you will need a QR code reader for your smartphone/tablet. There are many free readers available, depending on the smartphone/tablet you are using. We have supplied some suggestions below, but this is not an exhaustive list and you should only download software compatible with your device and operating system. We do not endorse any of the third-party products listed below and downloading them is at your own risk.

- for iPhone/iPad, search the App store for Qrafter
- for Android, search the Play store for QR Droid
- for Blackberry, search Blackberry World for QR Scanner Pro
- for Windows/Symbian, search the Store for Upcode

Once you have downloaded a QR code reader, simply open the reader app and use it to take a photo of the code. You will then see a menu of the free resources available for that topic.

### 1 Biological molecules

### 3 Cells

### 2 Enzymes

### 4 DNA and protein synthesis

5 The cell cycle

10 Mass transport in plants

6 The immune system

11 Genetic diversity

7 Gas exchange

12 Species and taxonomy

8 Digestion and absorption

13 Biodiversity within a community

9 Mass transport in animals

# Acknowledgements

The Publisher would like to thank the following for permission to reproduce copyright material:

**p.1** © evgenia sh – Fotolia; **p.3** © Yi Liu – Fotolia; **p.5** © Photo Insolite Realite/Science Photo Library; **p.16** © Martyn F. Chillmaid/Science Photo Library; **p.19** © fotografiche.eu – Fotolia; **p.28** © Patrik Stedrak – Fotolia; **p.36** © Sinclair Stammers/Science Photo Library; **p.37** *t* © Dr Jeremy Burgess/Science Photo Library, *b* © Science Photo Library; **p.39** *t* © CNRI/Science Photo Library, *b* © Susumu Nishinaga/Science Photo Library; **p.40** © A.R. Cavaliere, Dept of Biology, Gettysburg College, Gettysburg; **p.42** *t* © Biophoto Associates/Science Photo Library, *b* © Sinclair Stammers/Science Photo Library; **p.50** © Dr Gopal Murti/Science Photo Library; **p.51** © Moredun Animal Health Ltd/Science Photo Library; **p.56** *l* © Adrian T Sumner/Science Photo Library, *r* © A. Barrington Brown/Science Photo Library; **p.57** © Science Photo Library; **p.62** © Adrian T Sumner/Science Photo Library; **p.73** © M.I. Walker/Science Photo Library; **p.74** © Reuters/Corbis; **p.78** © Power And Syred/Science Photo Library; **p.80** © M.I. Walker/Science Photo Library; **p.82** © CNRI/Science Photo Library; **p.87** © yuuuu – Fotolia; **p.88** © Peter Menzel/Science Photo Library; **p.89** © Steve Gschmeissner/Science Photo Library; **p.90** © Dr Klaus Boller/Science Photo Library; **p.108** © Photographee.eu – Fotolia; **p.110** © Pascal Goetgheluck/Science Photo Library; **p.112** © Microfield Scientific Ltd/Science Photo Library; **p.115** © eAlisa – Fotolia; **p.116** © Wellcome Photo Library, Wellcome Images **p.119** *l* © James Steveson/Science Photo Library, *r* © CNRI/Science Photo Library; **p.124** *l* © James Steveson/Science Photo Library, *r* © James Stevenson/Science Photo Library; **p.130** © Eye Of Science/Science Photo Library; **p.131** © Power And Syred/Science Photo Library; **p.139** © Elena Schweitzer – Fotolia; **p.140** © Henri Bureau/Sygma/Corbis; **p.142** © Elena Schweitzer – Fotolia; **p.154** © Susumu Nishinaga/Science Photo Library; **p.156** © CNRI/Science Photo Library; **p.159** *t* © Mauro Fermariello/Science Photo Library, *b* © Andy Crump, TDR, WHO/Science Photo Library; **p.160** © D. Phillips/Science Photo Library; **p.165** © John Radcliffe Hospital/Science Photo Library; **p.167** © Dr Vim Jesudason; **p.168** *t* © Dr Vim Jesudason, *m* © Dr Vim Jesudason, *b* © Dr Vim Jesudason; **p. 169** *tl* © Dr Vim Jesudason, *tr* © Dr Vim Jesudason, *m* © Dr Vim Jesudason, *b* © Dr Vim Jesudason; **p.181** *l* © Garry DeLong – Fotolia, *m* © Steve Gschmeissner/Science Photo Library, *r* © Dr Keith Wheeler/Science Photo Library; **p.184** © enskanto – Fotolia; **p.185** © Dr Keith Wheeler/Science Photo Library; **p.189** © Brookhaven National Laboratory/Science Photo Library; **p.192** © hecke71 – Fotolia; **p.193** © L. Willatt, East Anglian Regional Genetics Service/Science Photo Library; **p.196** © Alila Medical Images/Alamy; **p.198** *l* © Piotr Filbrandt – Fotolia, *r* © daibui – Fotolia; **p.202** © CDC/Gilda L. Jones, CourtesyPublic Health Image Library; **p.210** © nickos68 – Fotolia; **p.211** *l* © Fotokon – Fotolia, *r* © fuujin – Fotolia; **p.213** *l* © ihervas – Fotolia, *r* © plrang – Fotolia; **p.214** *t* © birdiegal – Fotolia, *b* © Erni – Fotolia; **p.215** *l* © helmutvogler – Fotolia, *m* © Rich Lindie – Fotolia, *r* © nickos68 – Fotolia; **p.220** © Kajornyot – Fotolia; **p.225** © Jim Laws/Alamy; **p.231** © Mark Smith; **p.232** *t* © Mark Smith, *b* © andrewmroland – Fotolia; **p.235** © david crosbie – Fotolia; **p.239** © Science Photo Library; **p.246** © Bill Longcore/Science Photo Library; **p.247** © Professors P.M. Motta, S. Makabe & T. Naguro/Science Photo Library; **p.253** © Alexander Gospodinov – Fotolia; **p.255** *t* © Joao Inacio/Getty Images, *m* © Filip Ristevski – Fotolia, *bl* © Martyn F. Chillmaid/Science Photo Library, *br* © Science Photo Library; **p.260** *l* © Dr Keith Wheeler/Science Photo Library, *r* © Mark Smith; **p.261** *l* © Andrew Lambert Photography/Science Photo Library, *m* © Martyn F. Chillmaid/Science Photo Library, *r* © Andrew Lambert Photography/Science Photo Library; **p.262** © Pauline Calder; **p.263** © Alexander Gospodinov – Fotolia; **p.265** © Mira Arpe Bendevis, Dept. of Plant and Environmental Sciences, University of Copenhagen; **p.266** © ihervas – Fotolia.

*t* = top, *b* = bottom, *l* = left, *r* = right, *m* = middle

Every effort has been made to trace all copyright holders, but if any have been inadvertently overlooked, the Publisher will be pleased to make the necessary arrangements at the first opportunity.

The authors and the Publisher would also like to thank Martin Rowland for reviewing this book prior to publication.